Servsafe Study Guide:

DISCLAIMER: The information provided in this book is for educational purposes only and is not intended to serve as a substitute for professional advice, consultation, or training. The author and publisher have made every effort to ensure the accuracy and completeness of the information contained herein. However, they do not guarantee the accuracy or applicability of the content to any specific situation, and they assume no liability or responsibility for any errors, omissions, or consequences resulting from the use of the information provided.

Laws, regulations, and best practices related to food safety may vary between jurisdictions and are subject to change. Readers are advised to consult with appropriate authorities, experts, or legal counsel to ensure compliance with all applicable laws and regulations.

The certification exams mentioned in this book are administered by independent organizations, and the author and publisher have no affiliation with these organizations. Success in the exams is not guaranteed and depends on the individual's understanding, preparation, and effort.

By using this book, you agree that the author and publisher shall not be held liable for any damages or claims arising from the use or reliance on the information contained herein. It is the responsibility of the reader to assess the suitability of the content for their specific needs and circumstances.

PAGE 2 Introduction.
PAGE 7 Foodborne Illnesses and Microorganisms.
PAGE 14 Personal Hygiene and Employee Health.
PAGE 23 Cross-Contamination and Allergen Management.
PAGE 28 Cleaning, Sanitizing, and Pest Management.
PAGE 33 Facility Management and Design.
PAGE 39 Food Safety Regulations and Management Responsibilities.
PAGE 45-108 Practice Exam Section.

INTRODUCTION:

Welcome to this comprehensive study guide designed to help you achieve success on your ServSafe Food Manager and Certified Professional Food Manager (CPFM) exams. Embarking on this journey, you may have dreams of advancing your career, taking on new responsibilities, or simply ensuring the safety of your customers. We are here to encourage those dreams and provide you with the knowledge and tools necessary to excel in your food safety management journey.

We understand that preparing for these exams can be daunting, and you may have faced challenges in the past. However, this study guide is designed to ease your worries by breaking down complex concepts into easily digestible sections, providing you with clear explanations, practical examples, and engaging content. We are here to support you every step of the way and help you overcome any obstacles you may encounter.

Throughout this guide, we will cover a wide range of topics, including foodborne illnesses, personal hygiene, time and temperature control, cross-contamination, allergen management, cleaning and sanitizing, pest management, facility design, and food safety regulations. We will also provide exam preparation tips and test-taking strategies to alleviate your fears and ensure you feel confident going into the exams.

As you progress through the study guide, you will discover valuable insights and confirm your understanding of key food safety principles. We will also address common misconceptions and clarify any doubts you may have, so you can approach the exams with a solid foundation of knowledge and confidence.

While studying, it's natural to come across challenges or have questions about certain topics. Rest assured, this guide is designed to address those concerns and help you tackle any difficulties. Additionally, we've included practice exams with detailed explanations, giving you the opportunity to test your understanding and refine your skills before the actual exams.

Remember, the path to success is paved with determination, hard work, and a positive mindset. As you immerse yourself in this study guide, envision your future accomplishments and the positive impact you can make on your career and the lives of those you serve. We are here to encourage your dreams, support your learning journey, and celebrate your achievements.

So let's embark on this exciting adventure together, confident in the knowledge that you have the power to succeed and make a difference in the world of food safety management.

Food safety management is a critical aspect of the foodservice industry, as it not only ensures the health and well-being of customers but also fosters a successful and reputable business. As a professional in the food industry, you play a vital role in maintaining high standards of food safety and protecting the public from foodborne illnesses. Here are some key reasons why food safety management is essential:

1. Protect public health: Foodborne illnesses can lead to severe health consequences and, in some cases, even fatalities. By implementing effective food safety management practices, you can prevent the spread of harmful pathogens and protect the health of your customers.
2. Maintain a positive reputation: Food safety incidents can significantly damage a business's reputation, leading to a loss of customers, revenue, and credibility. By prioritizing food safety management, you demonstrate your commitment to customer well-being, fostering trust and loyalty among your clientele.
3. Comply with regulations: Food safety laws and regulations are in place to ensure the consistent delivery of safe and high-quality food products. Adhering to these guidelines not only keeps your business compliant but also helps you avoid fines, penalties, and potential closure.
4. Reduce food waste and costs: Effective food safety management practices, such as proper storage, handling, and temperature control, can minimize spoilage and waste. By reducing food waste, you can lower operational costs and contribute to a more sustainable food system.
5. Enhance employee morale and productivity: A well-trained and knowledgeable staff is essential for maintaining a safe and efficient foodservice environment. By investing in food safety management training, you empower your employees, leading to increased job satisfaction, productivity, and retention.
6. Improve business performance: A proactive approach to food safety management can help identify potential risks and hazards before they become problems. By addressing these issues promptly and effectively, you can ensure the smooth operation of your business and drive long-term success.

By prioritizing food safety management in your establishment, you not only protect the health of your customers but also create a strong foundation for a thriving business. This study guide will equip you with the knowledge and tools necessary to navigate food safety challenges, adhere to regulations, and excel as a responsible and successful foodservice professional.

Overview of ServSafe Manager and CPFM Exams

ServSafe Manager and Certified Professional Food Manager (CPFM) exams are designed to assess the knowledge and competency of foodservice professionals in managing food safety. These certifications demonstrate your commitment to upholding high standards of food safety and your ability to effectively manage foodservice operations. Below is a detailed overview of both exams.

ServSafe Manager Exam:

The ServSafe Manager Exam is administered by the National Restaurant Association (NRA) and focuses on the core principles of food safety management. The exam consists of 90 multiple-choice questions, with a passing score of 75% or higher. Test-takers have two hours to complete the exam. The main topics covered include:

Foodborne Illnesses and Microorganisms
Personal Hygiene and Employee Health
Time and Temperature Control for Safety (TCS) Foods
Cross-Contamination and Allergen Management
Cleaning, Sanitizing, and Pest Management
Facility Management and Design
Food Safety Regulations and Management Responsibilities
Certified Professional Food Manager (CPFM) Exam:

The CPFM Exam is offered by various accrediting organizations, such as Prometric, National Registry of Food Safety Professionals (NRFSP), and 360training. The exam format and passing score may vary depending on the organization, but the content is generally based on the FDA Food Code and other food safety regulations. The main topics covered in the CPFM exam include:

Foodborne Illnesses and Microorganisms
Personal Hygiene and Employee Health
Time and Temperature Control for Safety (TCS) Foods
Cross-Contamination and Allergen Management
Cleaning, Sanitizing, and Pest Management
Facility Management and Design
Food Safety Regulations and Management Responsibilities
Although the ServSafe Manager and CPFM exams are administered by different organizations, the content of both exams is quite similar, focusing on the core principles

of food safety management. To succeed in either exam, it is essential to be familiar with the latest food safety regulations, best practices, and requirements specific to your region or industry.

This study guide provides comprehensive coverage of the topics relevant to both the ServSafe Manager and CPFM exams. By following the guide and engaging with the practice questions and materials provided, you will be well-prepared to demonstrate your food safety management knowledge and skills on the exam day.

Book Structure:
This study guide is organized into logical sections and chapters, designed to facilitate your learning process and make it easier to grasp and retain essential food safety management concepts. The book structure is as follows:
1. Introduction
2. Foodborne Illnesses and Microorganisms
3. Personal Hygiene and Employee Health
4. Time and Temperature Control for Safety (TCS) Foods
5. Cross-Contamination and Allergen Management
6. Cleaning, Sanitizing, and Pest Management
7. Facility Management and Design
8. Food Safety Regulations and Management Responsibilities
9. Test-Taking Strategies and Tips
10. Practice Exams and Answer Explanations

Study Tips:
As you progress through this study guide, consider the following tips to maximize your learning and prepare effectively for the ServSafe Manager and CPFM exams:
1. Create a study schedule: Establish a regular study routine with dedicated time slots to review the material. Break down the content into smaller sections, focusing on one chapter or topic at a time.
2. Take notes and highlight key concepts: As you read, take notes or highlight essential information to reinforce your understanding and make it easier to review later.
3. Apply real-world examples: Relate the concepts you learn to real-life situations in your workplace or the foodservice industry. This will help solidify your understanding and make the material more relatable.
4. Review regularly: Periodically review previously studied material to reinforce your knowledge and enhance your long-term retention of the information.
5. Test your understanding with practice questions: Use the practice exams provided in this guide to test your knowledge and identify areas where you may

need to focus more attention. Review the answer explanations to understand why certain answers are correct or incorrect.
6. Teach others: Discussing the material with colleagues or explaining concepts to others can help reinforce your understanding and identify any gaps in your knowledge.
7. Stay up-to-date with food safety regulations: Regularly review the latest food safety regulations and best practices relevant to your region or industry, as these may change over time.
8. Maintain a positive mindset: Believe in your ability to succeed and approach your exam preparation with confidence and determination. Remember that practice and persistence will lead to success.

By following these study tips and engaging with the material in this study guide, you will be well-prepared to tackle the ServSafe Manager and CPFM exams and demonstrate your expertise in food safety management.

Foodborne Illnesses and Microorganisms

Foodborne illnesses are a significant public health concern, affecting millions of people worldwide each year. They are caused by the consumption of contaminated food or beverages, often due to the presence of harmful microorganisms. Understanding the different types of foodborne illnesses, their causes, and prevention methods is crucial for food safety management.

Types of Foodborne Illnesses:
1. Bacterial: Bacterial foodborne illnesses are caused by the ingestion of pathogenic bacteria, such as Salmonella, Escherichia coli (E. coli), Listeria monocytogenes, and Campylobacter. These bacteria can multiply rapidly in favorable conditions, such as warm temperatures and nutrient-rich environments.
2. Viral: Viruses, like Norovirus and Hepatitis A, can also cause foodborne illnesses. They are typically spread through person-to-person contact, contaminated surfaces, or the consumption of contaminated food or water.
3. Parasitic: Parasites, such as Giardia and Cryptosporidium, can contaminate food or water and cause gastrointestinal illness. Parasitic foodborne illnesses are often associated with consuming raw or undercooked meat, poultry, or seafood.
4. Fungal: Fungal foodborne illnesses are less common but can be caused by the ingestion of mold or yeast-contaminated foods. Some molds produce harmful toxins called mycotoxins, which can cause illness when ingested.

Prevention and Control Measures:
1. Time and temperature control: Maintaining proper temperature control throughout the food production process can significantly reduce the growth of harmful microorganisms. Keeping hot foods hot and cold foods cold minimizes the time spent in the "danger zone" (between 40°F and 140°F), where bacteria can multiply rapidly.
2. Proper food handling and storage: Following best practices for food handling and storage can help prevent cross-contamination and the spread of harmful microorganisms. This includes washing hands and surfaces regularly, separating raw and ready-to-eat foods, and properly storing food items.
3. Personal hygiene and employee health: Ensuring that food handlers practice good personal hygiene and are aware of the importance of reporting illnesses or symptoms is crucial in preventing the spread of foodborne pathogens.
4. Cooking and reheating: Cooking and reheating foods to the appropriate internal temperatures can help kill harmful microorganisms and reduce the risk of foodborne illnesses.

5. Cleaning and sanitizing: Regularly cleaning and sanitizing surfaces, utensils, and equipment can help eliminate harmful microorganisms and prevent cross-contamination.

Understanding the various types of foodborne illnesses and the microorganisms responsible for them is a critical aspect of food safety management. By implementing proper prevention and control measures, foodservice professionals can significantly reduce the risk of foodborne illnesses and protect the health and well-being of their customers.

Common Types of Foodborne Illnesses, Causes, and Symptoms:
1. Salmonella: Salmonella is a bacterial infection often linked to the consumption of raw or undercooked poultry, eggs, meat, and unpasteurized dairy products. Symptoms include diarrhea, fever, and abdominal cramps, typically appearing 12 to 72 hours after ingestion and lasting for 4 to 7 days.
2. E. coli: Escherichia coli (E. coli) infections can result from consuming contaminated ground beef, raw fruits and vegetables, and unpasteurized milk or juice. Symptoms include severe stomach cramps, diarrhea (often bloody), and vomiting, usually appearing 3 to 4 days after exposure and lasting for 5 to 7 days.
3. Listeria: Listeria monocytogenes is associated with consuming contaminated deli meats, soft cheeses, and raw sprouts. Symptoms include fever, muscle aches, diarrhea, and other gastrointestinal issues. In severe cases, Listeria can lead to meningitis or septicemia. Pregnant women, newborns, the elderly, and those with compromised immune systems are particularly at risk.
4. Campylobacter: Campylobacter infections typically result from consuming undercooked poultry, unpasteurized milk, or contaminated water. Symptoms include diarrhea (often bloody), fever, abdominal cramps, and vomiting, usually appearing 2 to 5 days after exposure and lasting for about a week.
5. Norovirus: Norovirus is a highly contagious viral infection, often spread through contaminated food, water, surfaces, or person-to-person contact. Common symptoms include diarrhea, vomiting, nausea, and stomach pain, appearing 12 to 48 hours after exposure and lasting for 1 to 3 days.

Importance of Understanding Foodborne Illnesses in Food Safety Management: Comprehending the causes, symptoms, and transmission routes of foodborne illnesses is crucial for food safety management, as it helps foodservice professionals take the necessary steps to prevent their occurrence. Prevention measures include:
1. Temperature control: Properly cooking, holding, cooling, and reheating foods can help prevent the growth of harmful microorganisms. Monitoring and maintaining appropriate temperatures throughout the food preparation process is essential.

2. Personal hygiene: Ensuring that employees follow good personal hygiene practices, such as handwashing, using gloves when appropriate, and not working when sick, can minimize the risk of transmitting foodborne pathogens.
3. Cross-contamination prevention: Separating raw and ready-to-eat foods, using designated cutting boards and utensils, and regularly sanitizing surfaces can help prevent the spread of harmful microorganisms between food items.
4. Proper food storage: Storing food at the correct temperatures and in appropriate containers can reduce the risk of contamination and growth of foodborne pathogens.
5. Employee training: Providing staff with proper training in food safety practices, including understanding the risks and symptoms of foodborne illnesses, helps ensure that foodservice operations prioritize safety and adhere to established guidelines.

By understanding the various types of foodborne illnesses and implementing the appropriate prevention measures, foodservice professionals can minimize the risk of foodborne outbreaks and protect the health and well-being of their customers.

Primary Sources and Transmission Routes of Foodborne Pathogens:
1. Bacteria: Bacterial pathogens like Salmonella, E. coli, Listeria, and Campylobacter are commonly found in raw or undercooked animal products such as meat, poultry, eggs, and unpasteurized dairy products. They can also be present in fruits, vegetables, and water. Bacteria can be transmitted through cross-contamination, improper storage, and inadequate cooking.
2. Viruses: Viral pathogens such as Norovirus and Hepatitis A are typically spread through person-to-person contact, touching contaminated surfaces, or consuming contaminated food or water. Viruses can be introduced to food by infected food handlers, contaminated raw ingredients, or unsanitary food preparation practices.
3. Parasites: Parasitic pathogens like Giardia and Cryptosporidium are often transmitted through contaminated water, raw or undercooked meat, poultry, or seafood. Contamination can occur when food comes into contact with fecal matter, either directly or indirectly, such as through contaminated water used for irrigation or processing.
4. Fungi: Fungal pathogens, including various molds and yeasts, can contaminate food through the presence of mold spores in the environment or by growing on spoiled or improperly stored food items. Some molds produce harmful mycotoxins, which can cause illness when ingested.

Steps to Minimize the Presence and Growth of Microorganisms:
1. Proper food handling and storage: Handle food with clean hands and utensils, separate raw and ready-to-eat foods, and store food items at appropriate temperatures to reduce the risk of cross-contamination and microbial growth.
2. Temperature control: Cook foods to the recommended internal temperatures to kill harmful microorganisms. Hold, cool, and reheat foods according to established guidelines to prevent the growth of pathogens.
3. Personal hygiene: Ensure that food handlers practice good personal hygiene, such as washing hands frequently, wearing gloves when necessary, and staying away from work when sick, to minimize the risk of transmitting pathogens.
4. Cleaning and sanitizing: Regularly clean and sanitize surfaces, utensils, and equipment to remove and kill harmful microorganisms. Pay particular attention to high-touch areas and places where raw and ready-to-eat foods are prepared.
5. Water safety: Use clean, potable water for food preparation and cleaning, and ensure that water sources are protected from contamination.
6. Pest control: Implement a pest control program to prevent pests from contaminating food and food preparation areas, as they can carry and transmit pathogens.
7. Employee training: Provide food safety training for staff to ensure they understand the risks associated with foodborne pathogens and follow established guidelines and best practices to minimize contamination and prevent the growth of harmful microorganisms.

By understanding the sources and transmission routes of foodborne pathogens and implementing these steps to minimize their presence and growth, foodservice professionals can significantly reduce the risk of foodborne illnesses and ensure a safe dining experience for their customers.

The "danger zone" is a term used in food safety management to describe the temperature range between 40°F (4°C) and 140°F (60°C). Within this range, harmful microorganisms such as bacteria can grow and multiply rapidly, increasing the risk of foodborne illnesses. The concept of the danger zone is crucial for food safety management, as it helps guide best practices in handling, storing, and preparing food. Controlling time and temperature plays a vital role in preventing the growth of harmful microorganisms and minimizing the risk of foodborne illnesses. Here are some key guidelines to follow:
1. Cook food thoroughly: Ensure that food is cooked to the recommended internal temperatures, which vary depending on the type of food. This will help kill any present microorganisms and reduce the risk of illness.

2. Hold hot and cold foods at proper temperatures: Hot foods should be held at or above 140°F (60°C), while cold foods should be kept at or below 40°F (4°C). This helps to inhibit the growth of harmful bacteria.
3. Cool food safely: When cooling cooked food, it's important to bring the temperature down from 140°F (60°C) to 70°F (21°C) within two hours and then down to 40°F (4°C) or below within the next four hours. Rapid cooling prevents bacteria from multiplying during the cooling process.
4. Reheat food properly: Reheating food should be done quickly and at a high temperature, usually at or above 165°F (74°C). This ensures that any remaining bacteria are destroyed and helps prevent the growth of harmful microorganisms.
5. Monitor time and temperature: Regularly check the temperature of food items during storage, preparation, and serving. Use a food thermometer to ensure accurate readings, and keep a log of temperatures to track adherence to food safety guidelines.

By understanding the significance of the danger zone in food safety management and implementing time and temperature control measures, you can effectively prevent the growth of harmful microorganisms, reduce the risk of foodborne illnesses, and ensure a safe and enjoyable dining experience.

Proper food handling and storage play a pivotal role in preventing foodborne illnesses, as they help minimize the risk of contamination and the growth of harmful microorganisms. Implementing best practices in a foodservice environment is essential to ensure the safety of food products and protect customers' health. Here are some specific examples of best practices for food handling and storage:

1. Hand hygiene: Food handlers should frequently wash their hands with warm water and soap, especially before and after handling food, after using the restroom, and after touching any surfaces that may be contaminated. This practice helps prevent the transfer of harmful microorganisms to food products.
2. Use of gloves: Gloves should be worn when handling ready-to-eat foods or when handling food with cuts or abrasions on the hands. Gloves should be changed frequently, especially when switching between tasks or when they become soiled.
3. Separating raw and ready-to-eat foods: To avoid cross-contamination, keep raw meats, poultry, and seafood separate from ready-to-eat foods, such as fruits, vegetables, and cooked items. Use separate cutting boards, utensils, and storage containers for raw and ready-to-eat foods.
4. Food storage hierarchy: Store food items in the correct order, based on their cooking temperatures, to prevent cross-contamination. For example, raw poultry

should be stored below raw beef or vegetables, as it has a higher cooking temperature requirement.
5. Temperature control during storage: Ensure that refrigerators and freezers are functioning correctly and maintain appropriate temperatures. Cold foods should be stored at or below 40°F (4°C), while frozen items should be kept at or below 0°F (-18°C).
6. FIFO (First In, First Out) method: Practice the FIFO method for food storage to minimize the risk of spoilage. Store new items behind older ones to ensure that the oldest products are used first.
7. Labeling and dating: Label and date all food items, particularly those that have been prepared in-house, to track the shelf life and ensure that they are used or discarded within the appropriate time frame.
8. Regular cleaning and sanitizing: Clean and sanitize food preparation surfaces, utensils, and equipment regularly to remove and kill harmful microorganisms. Pay special attention to high-touch areas, such as countertops, cutting boards, and refrigerator handles.

By implementing these best practices for food handling and storage in a foodservice environment, you can effectively minimize the risk of foodborne illnesses and create a safe dining experience for your customers.

Personal hygiene and employee health play a critical role in preventing foodborne illnesses, as food handlers can inadvertently introduce harmful microorganisms to food products if they do not adhere to proper hygiene practices or if they work while ill. Foodservice managers should take the following measures to ensure that their staff follow proper hygiene practices and understand the importance of reporting illnesses or symptoms:

1. Provide hygiene training: Conduct regular training sessions for staff members to educate them on the importance of personal hygiene and its impact on food safety. Teach them proper handwashing techniques, glove usage, and other personal hygiene practices.
2. Implement a handwashing policy: Establish a clear handwashing policy that requires staff to wash their hands frequently, especially before and after handling food, after using the restroom, and after touching potentially contaminated surfaces.
3. Enforce dress code and grooming standards: Implement a dress code that requires staff to wear clean uniforms, maintain trimmed fingernails, and restrain their hair to prevent contamination. Encourage the use of hairnets, hats, or other hair restraints when necessary.

4. Develop an illness reporting policy: Create a policy that requires employees to report any symptoms or illnesses that could potentially lead to foodborne illnesses. Encourage open communication and ensure that staff members understand the importance of reporting such conditions.
5. Monitor employee health: Regularly assess the health of your staff members and send home any employees displaying symptoms of illness, such as vomiting, diarrhea, or fever. Ensure that they do not return to work until they are symptom-free for the recommended period.
6. Provide resources for personal hygiene: Ensure that handwashing stations are readily available and stocked with soap, paper towels, and hand sanitizer. Place reminder signs near sinks to reinforce proper handwashing techniques.
7. Encourage the use of disposable tissues and no-touch waste receptacles: Make disposable tissues easily accessible for staff to use when they need to cough or sneeze. Provide no-touch waste receptacles to dispose of used tissues and other waste materials hygienically.
8. Regularly evaluate and reinforce hygiene practices: Continuously monitor staff adherence to personal hygiene practices and provide feedback or additional training when needed. Recognize and reward employees who consistently follow proper hygiene procedures to encourage compliance.

By taking these measures, foodservice managers can create a work environment that prioritizes personal hygiene and employee health, effectively minimizing the risk of foodborne illnesses and ensuring a safe dining experience for customers.

Personal Hygiene and Employee Health

Personal hygiene and employee health are crucial components of food safety in the foodservice industry. Ensuring that food handlers adhere to proper hygiene practices and maintain good health is essential for preventing the spread of foodborne illnesses and ensuring the safety of customers.

Key aspects of personal hygiene and employee health include:

1. Handwashing: Proper handwashing is vital for reducing the risk of contaminating food with harmful microorganisms. Food handlers should wash their hands frequently and thoroughly, using warm water and soap, especially before and after handling food, after using the restroom, and after touching potentially contaminated surfaces.
2. Dress code and grooming: Food handlers should wear clean uniforms, maintain proper grooming standards, and restrain their hair to minimize the risk of food contamination. This may include wearing hairnets, hats, or other hair restraints, as well as clean, closed-toe shoes.
3. Illness reporting and exclusion: Employees should be encouraged to report any symptoms or illnesses that could potentially lead to foodborne illnesses. Food handlers displaying symptoms of illness, such as vomiting, diarrhea, or fever, should be excluded or restricted from working with food until they are symptom-free for a specified period.
4. Training and reinforcement: Foodservice managers should provide regular training on personal hygiene practices and their importance in food safety. This includes monitoring staff adherence to hygiene policies and offering feedback or additional training as needed.
5. Personal habits: Food handlers should avoid engaging in habits that may lead to food contamination, such as smoking, eating, or drinking while working with food. Ensuring that these activities occur only in designated areas away from food preparation and service areas can help minimize the risk of contamination.

By prioritizing personal hygiene and employee health, foodservice establishments can effectively reduce the risk of foodborne illnesses and provide a safe dining experience for their customers.

Proper handwashing is a fundamental aspect of food safety, as it helps prevent the spread of foodborne illnesses by reducing the risk of transferring harmful microorganisms to food. The following steps outline the correct handwashing technique for food handlers:

1. Wet hands: Use warm, running water to wet your hands. The recommended temperature for handwashing is at least 100°F (38°C). Warm water is more effective at removing dirt and germs than cold water.
2. Apply soap: Apply a sufficient amount of soap to your hands, ensuring you have enough to create a good lather. Soap helps to lift dirt, grease, and microbes from your skin.
3. Lather and scrub: Vigorously rub your hands together to create a lather. Be thorough in your scrubbing, ensuring that you clean all areas of your hands, including the backs, between the fingers, under the fingernails, and around the wrists. This step should take at least 20 seconds to ensure you effectively remove any dirt and germs.
4. Rinse: Hold your hands under running water to rinse away the soap and any loosened dirt or germs. Make sure to rinse thoroughly, as any remaining soap residue can attract more dirt and bacteria.
5. Dry: Dry your hands using a clean paper towel or an air dryer. Damp hands can harbor bacteria and spread germs more easily than dry hands, so ensure they are completely dry before handling food.
6. Turn off the faucet: Use the paper towel, if available, to turn off the faucet to avoid recontaminating your hands.

By following these steps and practicing proper handwashing technique, food handlers can significantly reduce the risk of spreading foodborne illnesses, contributing to a safer foodservice environment for both employees and customers.

Maintaining a clean and appropriate dress code for foodservice employees is crucial in reducing the risk of food contamination and ensuring food safety. Adherence to proper attire guidelines is an important topic covered in the ServSafe Food Manager exam.

1. Hair restraints: Foodservice employees should wear hair restraints, such as hairnets, hats, or visors, to prevent loose hairs from falling into food. Hair can harbor bacteria and other contaminants, making hair restraints an essential measure in maintaining food safety.
2. Clean uniforms: Employees should wear clean uniforms or aprons to prevent cross-contamination between different food items and surfaces. Dirty clothing can transfer harmful microorganisms to food, so it is important to change uniforms or aprons regularly, especially if they become soiled during work.
3. Closed-toe shoes: Wearing closed-toe shoes in a foodservice environment helps protect employees' feet from potential injuries and prevents contaminants from being tracked into the kitchen. Closed-toe shoes also reduce the risk of foreign objects, such as dirt or debris, falling into food.

4. Jewelry restrictions: Jewelry can harbor bacteria and pose a risk of physical contamination if it falls into food. Foodservice employees should remove or minimize jewelry, particularly on their hands and wrists. The FDA Food Code advises against wearing jewelry, except for a plain ring, such as a wedding band.

By adhering to a clean and appropriate dress code, foodservice employees can significantly reduce the risk of food contamination, ensuring a safe dining experience for customers. Understanding these guidelines and their significance is vital for those preparing for the ServSafe Food Manager exam.

Excluding or restricting sick employees from working with food is a critical aspect of food safety management. This practice helps prevent the spread of foodborne illnesses, as sick employees may inadvertently contaminate food or surfaces with harmful pathogens. Understanding the importance of this measure and how to implement it is essential for those preparing for the ServSafe Food Manager exam.

1. Symptom-based exclusion: Employees should be excluded from working with food if they exhibit symptoms such as vomiting, diarrhea, jaundice, or sore throat with fever. These symptoms may indicate the presence of a foodborne illness, and excluding the employee from food handling activities is crucial in preventing the spread of infection.
2. Illness-based exclusion: Employees diagnosed with a foodborne illness, such as Salmonella, E. coli, or Hepatitis A, should be restricted from working with food. Managers must report these illnesses to their local health department to ensure proper follow-up and prevent further contamination.
3. Return-to-work guidelines: To determine when it is safe for an employee to return to work, managers should follow the guidelines provided by the FDA Food Code and local health authorities. These guidelines typically require that the employee be symptom-free for at least 24 hours or provide documentation from a medical professional stating that they are no longer contagious.
4. Alternative duties: While an employee is restricted from working with food, they may be assigned to other tasks that do not involve food handling, such as cleaning or administrative work. This approach helps maintain productivity while still prioritizing food safety.

Adhering to these guidelines ensures a safer foodservice environment and reduces the risk of foodborne illnesses. Familiarity with these concepts and their application is vital for those preparing for the ServSafe Food Manager exam.

Foodservice managers play a crucial role in promoting and enforcing personal hygiene practices among their staff to ensure food safety. Their responsibilities include providing

training, resources, and ongoing evaluation of hygiene compliance, all of which are essential components of the ServSafe Food Manager exam.
1. Training: Managers are responsible for educating staff on proper hygiene practices, such as handwashing techniques, appropriate dress code, and illness reporting procedures. Regular training sessions help reinforce these practices and keep employees up-to-date on food safety guidelines.
2. Resources: Managers should provide necessary resources to support personal hygiene practices, such as easily accessible handwashing stations with soap, water, and disposable towels. They should also supply appropriate attire, such as hair restraints and clean uniforms, to ensure staff adhere to dress code requirements.
3. Ongoing evaluation: Foodservice managers must consistently monitor and evaluate employees' adherence to personal hygiene practices. This may include observing handwashing frequency, checking for proper attire, and assessing overall cleanliness of the workspace.
4. Enforcement: Managers must establish a culture of accountability and take corrective action when necessary. This may involve addressing noncompliance issues, providing additional training, or implementing disciplinary measures for repeated violations.
5. Communication: Foodservice managers should maintain open lines of communication with their staff, encouraging employees to report any potential hygiene or health concerns. This fosters a collaborative environment where everyone is dedicated to maintaining food safety standards.

By actively promoting and enforcing personal hygiene practices, foodservice managers contribute to a safer dining experience for customers. Familiarity with these responsibilities and their implementation is vital for those preparing for the ServSafe Food Manager exam.

Personal habits such as smoking, eating, and drinking can have a significant impact on food safety in the workplace, as they can introduce contaminants into the food preparation and handling areas. To minimize the risk of contamination through these habits in a foodservice environment, it is essential to establish and follow strict guidelines, as required for the ServSafe Food Manager exam.
1. Designated areas: Create separate, designated areas for employees to smoke, eat, and drink, away from food preparation and storage areas. This helps prevent the spread of contaminants and ensures that personal habits do not interfere with food safety.
2. Handwashing: Require employees to wash their hands thoroughly after smoking, eating, or drinking, before returning to their workstations. Proper handwashing

is crucial to eliminate any potential contaminants that may have been introduced during these activities.
3. No-touch policies: Implement a no-touch policy for ready-to-eat foods, requiring employees to use utensils, gloves, or other barriers when handling such items. This minimizes the risk of contamination from hands that may have come into contact with personal items or surfaces during smoking, eating, or drinking.
4. Training and education: Provide ongoing training and education to employees on the importance of separating personal habits from food handling activities. Emphasize the significance of food safety and the potential consequences of not adhering to these guidelines.
5. Monitoring and enforcement: Regularly monitor employees' compliance with these policies and enforce them consistently. Address any violations promptly to maintain a strong food safety culture within the workplace.

By implementing these recommendations, foodservice managers can effectively minimize the risk of contamination from personal habits, ensuring a safe and healthy environment for customers. Understanding and applying these guidelines is crucial for those preparing for the ServSafe Food Manager exam.

The Time and Temperature Control for Safety (TCS) Foods chapter focuses on the importance of managing both time and temperature to ensure the safety of foods that are particularly susceptible to bacterial growth and foodborne illnesses. TCS foods are typically high in moisture, protein, and carbohydrates, providing an ideal environment for the growth of harmful microorganisms. Some common examples of TCS foods include meat, poultry, seafood, dairy products, eggs, cooked vegetables, and cooked grains.

Key points covered in this chapter include:
1. Temperature control: Proper temperature control is crucial for TCS foods to prevent the growth of harmful microorganisms. Maintain cold foods at or below 41°F (5°C) and hot foods at or above 135°F (57°C) during storage, preparation, and service. Avoid the "danger zone" (41°F to 135°F / 5°C to 57°C) where bacteria can grow rapidly.
2. Time control: Limit the amount of time TCS foods spend in the danger zone to minimize bacterial growth. Implement strict guidelines for food holding, cooling, and reheating to ensure food safety.
3. Cooking temperatures: Cook TCS foods to their minimum internal cooking temperatures to kill harmful microorganisms. Use a food thermometer to verify the internal temperature and ensure food safety.
4. Cooling and reheating: Follow proper cooling and reheating procedures for TCS foods to prevent bacterial growth. Cool cooked foods rapidly using appropriate

methods, such as ice baths or blast chillers, and reheat foods quickly to at least 165°F (74°C) before serving.
5. Hot and cold holding: Maintain appropriate holding temperatures for TCS foods during service, with hot foods at or above 135°F (57°C) and cold foods at or below 41°F (5°C).
6. Time as a public health control (TPHC): When using time instead of temperature to control bacterial growth, follow strict guidelines for discarding food that has been held at room temperature for a specified period, typically no more than four hours.

Understanding and implementing these time and temperature control principles is essential for maintaining the safety of TCS foods and reducing the risk of foodborne illnesses. This knowledge is crucial for those preparing for the ServSafe Food Manager exam and managing food safety in a foodservice environment.

Time and Temperature Control for Safety (TCS) foods have specific characteristics that make them more susceptible to bacterial growth, requiring strict time and temperature controls. TCS foods are typically moist, protein-rich, and have a neutral or slightly acidic pH, creating an ideal environment for bacteria to thrive.

Some common examples of TCS foods include:
1. Dairy products: Milk, cheese, and yogurt are protein-rich and can support bacterial growth if not handled properly.
2. Meat: Beef, pork, and poultry contain high levels of protein and moisture, which can facilitate the growth of harmful microorganisms.
3. Seafood: Fish and shellfish are also protein-rich and require proper time and temperature control to prevent bacterial growth.
4. Cooked vegetables: Cooked vegetables, such as potatoes or beans, can support the growth of bacteria if not cooled and stored correctly.
5. Cooked grains: Cooked rice, pasta, and other grains are starchy and moist, providing a suitable environment for bacterial growth if not handled correctly.

To reduce the risk of foodborne illness, it's crucial to maintain TCS foods at safe temperatures. Cold foods should be held at or below 41°F (5°C), while hot foods should be held at or above 135°F (57°C). This ensures that TCS foods remain outside the "danger zone" (41°F - 135°F or 5°C - 57°C), where harmful microorganisms grow most rapidly. Additionally, TCS foods should be cooked to their minimum internal cooking temperatures, which vary depending on the food item, to eliminate any pathogens that may be present.

The "danger zone" is a term used to describe the temperature range in which harmful microorganisms can grow rapidly, posing a risk for foodborne illnesses. For TCS foods, the danger zone lies between 41°F (5°C) and 135°F (57°C). Within this temperature range, bacteria, viruses, and other pathogens can multiply quickly, increasing the likelihood of food contamination.

Keeping TCS foods out of the danger zone is essential for maintaining food safety. Cold foods should be stored at or below 41°F (5°C), while hot foods should be held at or above 135°F (57°C). These temperature controls help minimize the growth of harmful microorganisms and reduce the risk of foodborne illnesses.

For example, imagine a restaurant that leaves a tray of cooked chicken out on the counter at room temperature (within the danger zone) for an extended period. Bacteria in the chicken can multiply rapidly, potentially causing illness if consumed. By ensuring the chicken is held at the appropriate hot holding temperature of 135°F (57°C) or higher, the risk of bacterial growth and subsequent foodborne illness can be significantly reduced.

Adhering to these temperature guidelines is a critical component of food safety and is essential for protecting customers and maintaining a healthy, compliant foodservice operation.

Cooking TCS foods to the proper internal temperatures is crucial for ensuring food safety and reducing the risk of foodborne illnesses. Using a food thermometer is an essential tool to verify that foods have reached their required minimum internal cooking temperatures, effectively killing harmful microorganisms.
Here are the minimum internal cooking temperatures for some key TCS food items:
1. Poultry (chicken, turkey, duck, and other fowl): Cook to an internal temperature of 165°F (74°C) for at least 15 seconds.
2. Ground meats (beef, pork, and other meats): Cook to an internal temperature of 155°F (68°C) for at least 15 seconds.
3. Seafood (fish, shellfish, and crustaceans): Cook to an internal temperature of 145°F (63°C) for at least 15 seconds.
4. Whole cuts of meat (beef, pork, veal, and lamb): Cook to an internal temperature of 145°F (63°C) and allow to rest for at least 3 minutes before serving.
5. Eggs: Cook to an internal temperature of 145°F (63°C) for at least 15 seconds, or 155°F (68°C) for dishes containing raw eggs.

For example, if a chef is preparing a chicken dish, they should use a food thermometer to ensure the internal temperature reaches 165°F (74°C) for at least 15 seconds. This will help guarantee the safety of the dish and protect customers from potential foodborne illnesses.

Remember to always clean and sanitize food thermometers between uses to avoid cross-contamination. By adhering to these guidelines and monitoring cooking temperatures with a food thermometer, foodservice establishments can uphold high food safety standards and reduce the risk of foodborne illnesses.

Proper cooling and reheating of TCS foods are essential to prevent bacterial growth and ensure food safety. Following the correct procedures helps to minimize the time foods spend in the danger zone, where harmful microorganisms can rapidly multiply.

Cooling TCS foods:
1. Cool foods from 135°F (57°C) to 70°F (21°C) within two hours. This first phase is critical because bacteria grow the fastest in this temperature range.
2. Continue cooling from 70°F (21°C) to 41°F (5°C) or below within the next four hours, completing the cooling process in a total of six hours.

To facilitate rapid cooling, use methods such as:
- Dividing large food portions into smaller containers.
- Using an ice water bath, placing the food container in a larger container filled with ice and water.
- Using commercial rapid cooling equipment, like a blast chiller.
- Stirring liquid foods periodically to distribute the heat evenly.

Reheating TCS foods:
Reheat TCS foods to an internal temperature of 165°F (74°C) for at least 15 seconds within two hours. This high temperature kills any harmful bacteria that may have grown during cooling or storage. When reheating, use equipment like ovens, stovetops, or microwaves that can rapidly heat the food and achieve the required temperature.

For example, when reheating a pot of chili, ensure it reaches 165°F (74°C) for at least 15 seconds using a food thermometer to verify the temperature. Avoid slow reheating methods, like using steam tables or chafing dishes, which can keep food in the danger zone for extended periods.

By adhering to these guidelines for cooling and reheating TCS foods, you can maintain high food safety standards and reduce the risk of foodborne illnesses.

Time as a Public Health Control (TPHC) is a food safety approach that allows TCS foods to be held without temperature control for a limited time, under specific conditions. TPHC is an alternative to traditional temperature control methods, and it helps prevent the growth of harmful microorganisms.

Circumstances for using TPHC:

TPHC is typically used when maintaining temperature control is challenging, such as during catered events, outdoor food festivals, or other situations where maintaining hot or cold holding temperatures is difficult.

Guidelines for implementing TPHC:

1. Obtain approval: Before using TPHC, obtain approval from the local regulatory authority, as some jurisdictions may have specific rules for its use.
2. Mark the time: Clearly mark the TCS food with the time it was removed from temperature control. It's essential to have a written procedure that indicates when the food was removed and when it must be discarded.
3. Time limits: TCS foods can be held without temperature control for up to four hours for cold foods (starting from 41°F / 5°C or below) and up to six hours for hot foods (starting from 135°F / 57°C or above). Once these time limits are reached, discard any remaining food.
4. Do not mix: Never mix fresh TCS food with food that's already under TPHC, as this can introduce bacteria into the new batch.
5. Discard food: TCS foods held at room temperature beyond the specified time limit must be discarded to prevent the growth of harmful microorganisms, reducing the risk of foodborne illnesses.

By understanding and implementing TPHC guidelines, foodservice professionals can manage TCS foods effectively in situations where maintaining traditional temperature control is challenging. This approach helps maintain high food safety standards while providing flexibility in various foodservice environments.

Cross-Contamination and Allergen Management

The Cross-Contamination and Allergen Management chapter focuses on the prevention of foodborne illnesses caused by the transfer of harmful substances, bacteria, or allergens between different food items, surfaces, and equipment. This chapter highlights the importance of understanding and implementing effective practices to minimize the risks associated with cross-contamination and allergen exposure in foodservice establishments. Key topics covered in this chapter include:

1. Cross-contamination: This occurs when harmful microorganisms or substances transfer from one food item or surface to another, potentially leading to foodborne illnesses. Common sources of cross-contamination include raw meats, poultry, seafood, and unwashed produce. Proper food handling, storage, and preparation techniques are essential to prevent cross-contamination.
2. Allergen management: Food allergies affect millions of people worldwide, and allergen exposure can cause mild to severe reactions, including life-threatening anaphylaxis. Foodservice establishments must be vigilant in managing allergens, including proper labeling, storage, preparation, and service of allergen-containing foods. Staff training and clear communication with customers are vital in ensuring allergen safety.
3. Personal hygiene: Food handlers must follow strict personal hygiene practices, such as handwashing, wearing gloves, and avoiding bare-hand contact with ready-to-eat foods, to prevent cross-contamination and allergen transfer.
4. Cleaning and sanitizing: Regular cleaning and sanitizing of surfaces, utensils, and equipment are essential in reducing the risk of cross-contamination and allergen exposure. A thorough cleaning process removes food particles and debris, while sanitizing helps eliminate harmful microorganisms.
5. Separation of foods: Foodservice establishments must store and prepare different types of foods separately to minimize cross-contamination risks. This includes using separate cutting boards, utensils, and storage areas for raw meats, poultry, seafood, and allergen-containing foods.

By understanding and implementing the principles and practices outlined in the Cross-Contamination and Allergen Management chapter, foodservice professionals can help ensure a safe and enjoyable dining experience for all customers, including those with food allergies or sensitivities.

Cross-contamination is a significant concern in food safety, as it can lead to the spread of harmful pathogens and allergens. There are two main types of cross-contamination: direct and indirect.

Direct cross-contamination occurs when harmful substances or microorganisms transfer directly from one food item to another. For instance, imagine preparing raw chicken on a cutting board, and then immediately chopping vegetables on the same board without cleaning it. The raw chicken juices, potentially containing pathogens like Salmonella, can contaminate the vegetables, posing a risk to consumers.

Indirect cross-contamination happens when contaminants transfer between food items through an intermediary, such as utensils, equipment, or even human hands. For example, if a chef uses a knife to cut raw meat and then uses the same knife to slice a tomato without washing it, the tomato could be contaminated with bacteria from the meat.

The consequences of cross-contamination can be severe, particularly for individuals with compromised immune systems, the elderly, pregnant women, and young children. Cross-contamination can lead to foodborne illnesses, which may cause symptoms ranging from mild discomfort to severe dehydration, hospitalization, or even death in extreme cases.

To mitigate the risks of cross-contamination, foodservice establishments should adhere to strict food safety protocols, including proper handwashing, using separate cutting boards and utensils for different food types, and ensuring regular cleaning and sanitizing of all surfaces and equipment. By taking these precautions, the risk of cross-contamination can be significantly reduced, ensuring a safer dining experience for all patrons.

The FDA identifies eight major food allergens, which account for around 90% of all food allergy reactions. These allergens include:
1. Milk
2. Eggs
3. Fish
4. Crustacean shellfish (e.g., shrimp, crab, lobster)
5. Tree nuts (e.g., almonds, walnuts, pecans)
6. Peanuts
7. Wheat
8. Soybeans

Allergic reactions to these foods can vary in severity, ranging from mild symptoms like itching, hives, and digestive discomfort, to life-threatening anaphylaxis, which may involve difficulty breathing, rapid or weak pulse, and a sudden drop in blood pressure.

Foodservice establishments play a critical role in preventing allergen cross-contact and ensuring accurate communication with customers about allergens present in menu items. Key responsibilities include:
1. Staff training: Ensure all staff members, from the kitchen to the front-of-house, are knowledgeable about allergens, their potential effects, and the steps to prevent cross-contact.
2. Ingredient tracking: Maintain detailed, up-to-date information on the ingredients used in each menu item, including the presence of allergens.
3. Separate preparation: Use designated cutting boards, utensils, and equipment for allergen-containing foods to prevent cross-contact.
4. Cleaning and sanitizing: Regularly clean and sanitize surfaces, equipment, and utensils to minimize the risk of allergen cross-contact.
5. Clear communication: Accurately convey allergen information to customers, both through the menu and via staff interactions, ensuring they have the necessary information to make informed choices.

By implementing these practices, foodservice establishments can minimize the risk of allergen cross-contact and protect the well-being of their customers with food allergies.

Proper handwashing and personal hygiene practices are crucial in preventing cross-contamination and allergen cross-contact in foodservice settings. These practices reduce the risk of harmful microorganisms and allergens being transferred from hands, clothing, or surfaces to food, ensuring the safety of customers.

Effective handwashing guidelines include:
1. Wetting hands with warm water.
2. Applying soap and lathering for at least 20 seconds, covering all surfaces of the hands and fingers, including under the nails.
3. Rinsing thoroughly with warm water.
4. Drying hands with a single-use paper towel or air dryer.

In addition to regular handwashing, staff should wash their hands in specific situations, such as:
- Before starting work.
- After using the restroom.
- After touching their face, hair, or body.
- After handling raw animal products or allergen-containing ingredients.
- After handling trash, cleaning chemicals, or dirty equipment.

The use of gloves is another essential practice to prevent cross-contamination and allergen cross-contact. Here are some guidelines and best practices for using gloves in a foodservice setting:
1. Always wash hands before putting on gloves.

2. Choose gloves that fit well and are made of appropriate materials for the task.
3. Change gloves frequently, especially when switching tasks or handling different types of foods.
4. Discard gloves if they become torn, contaminated, or used for an extended period.
5. Never use gloves as a substitute for handwashing.

By implementing proper handwashing and personal hygiene practices, foodservice establishments can minimize the risk of cross-contamination and allergen cross-contact, ensuring a safer dining experience for customers.

Cleaning and sanitizing play a critical role in minimizing cross-contamination and allergen exposure risks in foodservice environments. Properly cleaned and sanitized surfaces, cutting boards, utensils, and equipment prevent the transfer of harmful microorganisms and allergens between different foods, ensuring customer safety.
To effectively clean and sanitize cutting boards, utensils, surfaces, and equipment, follow these steps:

1. Cleaning: Begin by removing any visible debris from the item or surface using a scraper, brush, or cloth. Then, apply a detergent solution to break down grease and food particles. Scrub thoroughly, ensuring all surfaces are covered, and rinse with clean water to remove any detergent residue.
2. Sanitizing: After cleaning, apply a sanitizing solution to eliminate any remaining microorganisms. Sanitizing solutions can be chemical-based (such as chlorine, iodine, or quaternary ammonium compounds) or heat-based (using hot water or steam). Follow the manufacturer's instructions for the correct concentration, contact time, and temperature when using chemical sanitizers. For heat sanitizing, immerse items in water at a temperature of at least 171°F (77°C) for a minimum of 30 seconds.
3. Air-drying: Allow all items and surfaces to air-dry completely. Do not use towels or cloths for drying, as they can reintroduce contaminants.

For cutting boards, it's essential to use separate boards for different food types, such as raw meats, cooked meats, and fresh produce. This practice helps prevent cross-contamination. Additionally, replace cutting boards that are excessively worn or have deep grooves, as they can harbor bacteria and allergens.
Regularly clean and sanitize high-touch surfaces, such as handles, faucets, and switches, to reduce the risk of cross-contamination. Establish a cleaning and sanitizing schedule for your establishment, ensuring all areas are consistently maintained.
By following proper cleaning and sanitizing techniques, foodservice establishments can minimize cross-contamination and allergen exposure risks, promoting a safe and healthy dining experience for customers.

Separate storage and preparation areas for different types of foods are essential to minimize cross-contamination and allergen exposure risks in a foodservice environment. By segregating raw meats, poultry, seafood, and allergen-containing foods, you can prevent the transfer of harmful microorganisms and allergens between food items. Here are some strategies for implementing effective separation practices in a foodservice environment:

1. Designated storage spaces: Allocate separate, clearly marked storage spaces for raw meats, poultry, seafood, produce, and allergen-containing foods. Store raw animal products below ready-to-eat foods and produce to prevent drips and spills from contaminating other items.
2. Color-coded equipment: Use color-coded cutting boards, utensils, and containers to distinguish between different food types. For example, assign specific colors for raw meats, cooked meats, produce, and allergen-containing foods. This practice helps staff members quickly identify the appropriate tools for each food type and reduces the risk of cross-contamination.
3. Physical barriers: In the kitchen, establish distinct preparation areas for raw meats, poultry, seafood, produce, and allergen-containing foods. Use physical barriers, such as counters, shelves, or partitions, to separate these areas and prevent cross-contamination.
4. Staff training: Educate your staff on the importance of food segregation and provide clear guidelines for handling different food types. Emphasize the consequences of cross-contamination and allergen exposure and ensure all team members understand their responsibilities in maintaining a safe foodservice environment.
5. Proper cleaning and sanitizing: Regularly clean and sanitize all surfaces, cutting boards, utensils, and equipment in accordance with the guidelines discussed earlier. This practice helps to eliminate any residual contaminants and further reduces the risk of cross-contamination.

By implementing these strategies, foodservice establishments can effectively minimize cross-contamination and allergen exposure risks, ensuring a safe dining experience for customers and maintaining a high standard of food safety.

Cleaning, Sanitizing, and Pest Management

The Cleaning, Sanitizing, and Pest Management chapter focuses on essential practices for maintaining a clean and safe foodservice environment. It covers the following key areas:

1. Cleaning and sanitizing: This section outlines the importance of proper cleaning and sanitizing procedures to eliminate foodborne pathogens and allergens from surfaces, utensils, and equipment. It explains the difference between cleaning (removing visible dirt and debris) and sanitizing (reducing the number of microorganisms to safe levels). The chapter also covers the appropriate use of detergents, sanitizers, and the correct techniques for cleaning and sanitizing various items in a foodservice establishment.
2. Cleaning schedules: The chapter emphasizes the importance of establishing and adhering to a regular cleaning schedule. This ensures that all areas of the foodservice establishment, including surfaces, equipment, utensils, and storage areas, are consistently maintained to prevent the buildup of harmful microorganisms and allergens.
3. Pest management: Effective pest management is crucial for food safety, as pests can carry pathogens and contaminate food, surfaces, and equipment. This section outlines integrated pest management strategies that include preventive measures, such as proper sanitation, structural maintenance, and trash management, along with corrective actions, like using approved pesticides or partnering with professional pest control services.
4. Monitoring and documentation: The chapter highlights the importance of regularly monitoring cleaning, sanitizing, and pest management practices. It discusses the need for maintaining accurate records of these activities to ensure compliance with food safety regulations and demonstrate a commitment to maintaining a safe foodservice environment.

Overall, the Cleaning, Sanitizing, and Pest Management chapter provides essential guidance on maintaining a clean and safe foodservice environment, which is crucial for preventing foodborne illness and ensuring a high standard of food safety.

Cleaning and sanitizing are two distinct processes that play a crucial role in maintaining a safe foodservice environment. While they are often used interchangeably, it's essential to understand the differences between them and their specific applications.

Cleaning refers to the process of removing visible dirt, debris, and organic matter from surfaces and equipment. This is typically done using water, detergents, and scrubbing tools. Cleaning helps reduce the presence of microorganisms but does not eliminate them completely. Examples of cleaning agents include dish soap, degreasers, and all-purpose cleaners.

Sanitizing, on the other hand, is the process of reducing the number of harmful microorganisms on a surface to a safe level. Sanitizing does not remove visible dirt or debris; it focuses on eliminating bacteria, viruses, and other pathogens. Sanitizers are applied after cleaning, and examples include chlorine bleach, quaternary ammonium compounds (QACs), and iodine-based sanitizers.

Both cleaning and sanitizing are vital in a foodservice environment because they help prevent cross-contamination, the spread of foodborne illnesses, and maintain overall food safety. It's essential to follow the appropriate procedures for each process and use the correct cleaning and sanitizing agents for specific surfaces and equipment.

For instance, chlorine bleach can be used as a sanitizer for cutting boards, utensils, and countertops, while QACs are commonly used for sanitizing food contact surfaces and equipment. It's important to follow the manufacturer's instructions for the correct concentration, contact time, and application method when using cleaning and sanitizing agents. By understanding and implementing both cleaning and sanitizing processes effectively, foodservice establishments can help ensure a safe and healthy environment for staff and customers.

Creating a comprehensive cleaning schedule is essential for maintaining a clean and safe foodservice environment. A well-organized cleaning schedule should include the frequency of cleaning tasks, areas and equipment to be cleaned, and staff responsibilities.
1. Frequency of cleaning tasks: Cleaning tasks can be categorized as daily, weekly, or monthly, depending on the nature of the task and its impact on food safety. For instance, daily tasks may include sweeping and mopping floors, sanitizing cutting boards and utensils, and wiping down countertops. Weekly tasks might involve deep-cleaning appliances, scrubbing walls and ceilings, or cleaning storage areas. Monthly tasks could include checking and cleaning grease traps or inspecting equipment for wear and tear.
2. Areas and equipment to be cleaned: A cleaning schedule should address all areas of the foodservice establishment, including the kitchen, storage areas, dining areas, restrooms, and outdoor spaces. It should also outline the cleaning procedures for specific equipment, such as ovens, grills, refrigerators, and

dishwashing machines. Don't forget about less obvious items, like trash bins, sinks, and light fixtures.
   3. Staff responsibilities: Assigning specific cleaning tasks to individual staff members helps ensure accountability and thoroughness. Clearly communicate who is responsible for each task and provide proper training on the correct cleaning and sanitizing procedures. Encourage staff to take ownership of their assigned tasks and monitor their performance to ensure compliance with the cleaning schedule.

By developing a comprehensive cleaning schedule and ensuring staff members understand their responsibilities, foodservice establishments can maintain a clean and safe environment. Regular cleaning helps prevent cross-contamination, reduce the risk of foodborne illness, and promote a positive image for customers.

Integrated Pest Management (IPM) is a systematic approach to controlling pests in a foodservice setting by employing preventive measures, monitoring techniques, and corrective actions. This strategy aims to minimize the use of chemicals and emphasize environmentally friendly methods to ensure food safety and maintain a clean environment.
   1. Preventive measures: The first line of defense against pests is prevention. Seal any openings or cracks in the building's exterior, install screens on windows and vents, and maintain proper door seals. Regularly clean and sanitize all areas, paying particular attention to floors, walls, and hidden spaces where pests may hide. Store food in tightly sealed containers and promptly remove any spoiled or contaminated items. Dispose of trash properly and maintain clean, well-organized storage areas.
   2. Monitoring techniques: Regularly inspect the foodservice establishment for signs of pests, such as droppings, nests, or damaged packaging. Establish a pest sighting log for staff to report any pest activity, and periodically review this log to identify patterns or problem areas. Use traps or sticky monitoring devices strategically placed throughout the facility to help detect and track pest activity.
   3. Corrective actions: If pests are detected, take immediate action to address the issue. Consult with a licensed pest control professional to identify the specific pest and determine the most effective treatment strategy. Implement targeted measures to eliminate the infestation, such as using traps, baits, or insecticides, as recommended by the pest control expert. Always follow label instructions and applicable regulations when using chemical treatments.

In summary, an effective IPM program in a foodservice setting involves proactive preventive measures, regular monitoring, and prompt corrective actions when pests are detected. By implementing these principles, foodservice establishments can successfully control pests and maintain a clean, safe environment that protects food, surfaces, and equipment from contamination.

Proper storage, handling, and disposal of cleaning and sanitizing chemicals are crucial in a foodservice establishment to ensure the safety of both staff and customers, as well as compliance with relevant regulations.

1. Safe storage: Chemicals should be stored separately from food, utensils, and other kitchen items to prevent cross-contamination. They should be kept in their original containers with clear labels, or if transferred, the new containers should be labeled with the chemical's name and hazard information. Store chemicals in a cool, dry, and well-ventilated area, away from direct sunlight or heat sources, to avoid chemical reactions or degradation.
2. Safe handling: Staff should be trained on the proper handling of chemicals, including using personal protective equipment (PPE) such as gloves, goggles, and aprons when necessary. They should also be educated on how to read and understand Safety Data Sheets (SDS) for each chemical, which provide essential information on the chemical's hazards, protective measures, and emergency procedures. Mixing chemicals should be avoided unless specified by the manufacturer, as some combinations can create toxic gases or cause other dangerous reactions.
3. Safe disposal: Dispose of chemicals according to the manufacturer's instructions and local regulations, as improper disposal can harm the environment and pose risks to public health. Do not pour chemicals down drains or into trash containers without appropriate treatment. In some cases, chemical waste may need to be collected and transported by a licensed waste disposal company for proper treatment or disposal.

By adhering to these guidelines for the storage, handling, and disposal of cleaning and sanitizing chemicals, foodservice establishments can protect the safety of staff and customers, maintain a clean environment, and comply with relevant regulations. Implementing these practices also contributes to a professional and responsible approach to food safety and overall business operations.

Monitoring and documentation play a crucial role in cleaning, sanitizing, and pest management practices within foodservice establishments. They serve to demonstrate adherence to food safety regulations and standards while maintaining a consistently safe and hygienic environment.

1. Monitoring: Regularly inspecting and evaluating the effectiveness of cleaning, sanitizing, and pest control efforts helps identify areas that require attention or improvement. Monitoring also involves checking the proper use and storage of

chemicals, ensuring that cleaning schedules are followed, and verifying that pest management measures are effective. Tools such as checklists, visual inspections, and test kits for sanitizer concentration can be employed to facilitate monitoring.

2. Documentation: Keeping accurate and up-to-date records of cleaning, sanitizing, and pest management activities serves several purposes:

a. Regulatory compliance: Documentation demonstrates that the establishment follows food safety regulations and guidelines, which can be crucial during inspections or audits by health departments or other regulatory agencies.

b. Traceability and accountability: Records help track the specific actions taken, the individuals responsible, and any corrective measures implemented. This information can be useful for employee training, performance evaluation, or investigating food safety incidents.

c. Continuous improvement: Analyzing records can reveal patterns or trends that indicate areas where procedures need to be modified, equipment upgraded, or staff retrained to enhance food safety and sanitation.

Types of records that should be maintained include:

i. Cleaning and sanitizing schedules and logs: These detail the frequency, methods, and specific areas or equipment cleaned, as well as the chemicals used, their concentrations, and the responsible staff members.

ii. Pest management logs: These provide information on pest sightings, preventive measures (e.g., structural repairs or employee training), monitoring methods (e.g., traps or inspections), and corrective actions (e.g., pesticide applications or sanitation improvements).

iii. Chemical inventory and Safety Data Sheets (SDS): An up-to-date inventory of cleaning and sanitizing chemicals, along with their SDS, should be readily available to staff for reference and training purposes.

By closely monitoring and documenting cleaning, sanitizing, and pest management practices, foodservice establishments can maintain a safe and hygienic environment, demonstrate compliance with food safety regulations, and continuously improve their operations to safeguard the health and well-being of both staff and customers.

Facility Management and Design

The Facility Management and Design chapter provides an overview of the essential elements that contribute to creating and maintaining a safe and efficient foodservice environment. The chapter focuses on various aspects of facility design, layout, equipment selection, and maintenance, emphasizing their significance in ensuring food safety, employee safety, and operational efficiency. Key topics covered in this chapter include:

1. Facility design and layout: The chapter discusses the importance of proper design and layout of a foodservice establishment, which can directly impact food safety, employee productivity, and customer satisfaction. Factors to consider include appropriate spacing, workflow, and separation of different areas (such as food preparation, storage, and dishwashing) to minimize cross-contamination and maintain a clean environment.
2. Flooring, walls, and ceilings: The selection of materials for floors, walls, and ceilings is essential for maintaining cleanliness and durability. The chapter covers the characteristics of suitable materials, such as being smooth, non-absorbent, and easy to clean, as well as guidelines for installation and maintenance.
3. Lighting and ventilation: Adequate lighting and ventilation play a critical role in creating a safe and comfortable work environment. The chapter outlines the recommended levels of illumination for various areas within a foodservice establishment and the importance of proper ventilation systems to control odors, humidity, and airborne contaminants.
4. Equipment selection and maintenance: Choosing the right equipment and keeping it well-maintained is vital for food safety and operational efficiency. The chapter discusses factors to consider when selecting equipment, such as compliance with relevant safety standards, ease of cleaning, and energy efficiency. Additionally, it highlights the importance of regular maintenance, cleaning, and calibration to ensure equipment functions optimally.
5. Water supply and plumbing: A safe and reliable water supply is crucial for food preparation, cleaning, and sanitation. The chapter covers the requirements for potable water sources, water temperature controls, backflow prevention, and proper drainage systems to prevent contamination.
6. Waste disposal and recycling: Proper waste management is essential for maintaining a sanitary foodservice environment. The chapter outlines strategies for efficient waste segregation, storage, and disposal, as well as methods for implementing recycling and composting programs.

7. Employee facilities: Providing suitable facilities for employees, such as restrooms, locker rooms, and break areas, is important for promoting hygiene and comfort. The chapter discusses the requirements for these facilities, including their design, location, and maintenance.

By understanding and implementing the principles of facility management and design, foodservice operators can create a safe, efficient, and pleasant environment that contributes to the overall success of the establishment while ensuring compliance with food safety regulations.

When designing a foodservice establishment's layout, there are several critical factors to consider, as a well-thought-out design can ensure the smooth operation and overall food safety. Let's dive into some of these essential aspects.

1. Separation of areas: Allocate distinct spaces for food preparation, cooking, storage, dishwashing, and waste disposal. This separation helps avoid cross-contamination and maintain an organized work environment. Designate specific areas for handling raw meats, poultry, and seafood, as well as allergen-containing foods, to further minimize contamination risks.
2. Workflow efficiency: Arrange workstations, equipment, and storage areas in a logical sequence that mirrors the flow of food through the establishment. This can be achieved through a linear, parallel, or island-style layout. By designing the space for optimal efficiency, you can reduce the time employees spend moving between tasks and minimize the risk of errors.
3. Minimizing cross-contamination risks: Incorporate features such as splash guards, sneeze guards, and foot-operated sinks to reduce the spread of contaminants. Ensure handwashing stations are easily accessible and strategically placed throughout the facility, particularly near high-risk areas like raw food handling stations.
4. Space and ergonomics: Allocate enough space for employees to move around comfortably and perform tasks without bumping into each other or equipment. Consider the height and reach of countertops and storage areas, ensuring they are ergonomically designed to minimize strain and injuries.
5. Flexibility and adaptability: Design the layout to accommodate changes in menu offerings, equipment upgrades, and staff needs. This can be achieved by incorporating modular or mobile workstations, adjustable shelving, and easily reconfigurable spaces.

By carefully considering these key factors, you can create a foodservice establishment layout that promotes efficiency, safety, and a pleasant working environment for staff.

The selection of appropriate materials for floors, walls, and ceilings in a foodservice facility is crucial for maintaining cleanliness, ensuring durability, and adhering to food safety regulations. Here are some key characteristics to consider when choosing materials for each surface:

1. Floors: The ideal flooring material should be durable, non-absorbent, easy to clean, and slip-resistant. Common choices include epoxy resin, rubber, and quarry or ceramic tiles. Additionally, coved or curved junctions between the floor and the wall can prevent dirt and debris accumulation, making cleaning more effective.
2. Walls: Wall surfaces should be smooth, non-absorbent, and resistant to chipping or flaking. Materials such as stainless steel, fiberglass-reinforced plastic, and ceramic tiles are popular choices. Choose light colors to enhance visibility and facilitate the detection of dirt and grime. Also, consider incorporating impact-resistant materials in high-traffic areas to protect walls from damage.
3. Ceilings: Ceiling materials should be smooth, non-absorbent, and easy to clean. Drop ceilings, although aesthetically pleasing, should be avoided in food preparation areas, as they can harbor pests and make cleaning difficult. Materials like fiberglass-reinforced plastic or vinyl-coated gypsum panels are suitable options.
4. Joints and seams: Ensure that joints and seams in floors, walls, and ceilings are sealed with materials like silicone sealants or epoxy grouts. Sealed joints prevent the accumulation of dirt, moisture, and pests, promoting a more hygienic environment.
5. Compliance with regulations: Select materials that comply with local and national food safety regulations, which may include specific requirements for surface materials, finishes, and maintenance procedures.

By choosing appropriate materials for floors, walls, and ceilings in a foodservice facility, you can create an environment that is not only visually appealing but also easy to clean, durable, and compliant with food safety standards.

Proper lighting and ventilation play significant roles in maintaining a safe and efficient foodservice environment. Here's an overview of their importance and recommendations for various areas:

1. Lighting: Adequate illumination is crucial for ensuring food safety, cleanliness, and worker productivity. Different areas of a foodservice facility require specific lighting levels:

- Food preparation areas: 50 to 75 foot-candles (538 to 807 lux) to ensure proper visibility for tasks like cutting, chopping, and measuring.

- Dishwashing areas: 20 to 50 foot-candles (215 to 538 lux) to facilitate the identification of food residue on dishes and utensils.
- Storage areas: 10 to 30 foot-candles (108 to 323 lux) for general visibility and inventory management.
- Dining areas: 5 to 20 foot-candles (54 to 215 lux) to create a comfortable and inviting atmosphere for customers.

Ensure lighting fixtures are protected with shatter-resistant covers to prevent glass contamination in case of breakage.

2. Ventilation: Effective ventilation systems are essential for controlling odors, humidity, and airborne contaminants, promoting a comfortable environment for staff and customers alike.
   - Exhaust hoods: Install exhaust hoods over cooking equipment to remove heat, smoke, and grease-laden vapors. Properly designed hoods capture and filter contaminants before discharging them outside the building.
   - Make-up air: Provide sufficient make-up air to replace the air removed by exhaust hoods. This prevents negative air pressure, which can cause backdrafts and hinder the effectiveness of exhaust systems.
   - Air circulation: Ensure adequate air circulation to maintain a comfortable working environment and prevent the growth of mold and mildew. Air movement should be between 30 and 50 cubic feet per minute (CFM) per person in occupied spaces.
   - Air filtration: Use air filters with appropriate Minimum Efficiency Reporting Value (MERV) ratings to reduce airborne contaminants, such as dust, allergens, and mold spores. Replace filters as recommended by the manufacturer.

By implementing proper lighting and ventilation in a foodservice facility, you can enhance food safety, maintain cleanliness, boost employee productivity, and create a comfortable atmosphere for both staff and customers.

Selecting and maintaining equipment for a foodservice establishment requires careful consideration of various factors to ensure safety, efficiency, and compliance with regulations. Let's explore these factors in more detail:

1. Compliance with safety standards: Choose equipment that meets established safety standards, such as those set by the National Sanitation Foundation (NSF) or Underwriters Laboratories (UL). This ensures that the equipment is designed to minimize hazards and promote food safety.
2. Ease of cleaning: Opt for equipment that is easy to clean and sanitize, with smooth, nonporous surfaces and minimal crevices. This helps prevent the buildup of food debris and bacteria, contributing to a cleaner and safer food preparation environment.

3. Energy efficiency: Energy-efficient equipment not only reduces operational costs but also contributes to environmental sustainability. Look for appliances with the ENERGY STAR® certification, which indicates they meet strict energy efficiency guidelines set by the U.S. Environmental Protection Agency (EPA).
4. Regular maintenance and calibration: Implement a routine maintenance schedule to keep equipment in optimal working condition. Regular maintenance can help prevent breakdowns, extend the life of the equipment, and ensure consistent food quality. Calibration is especially important for temperature-controlled devices, such as ovens and refrigerators, to maintain accurate temperature readings and prevent food spoilage.
5. Ergonomics and workflow: Select equipment that facilitates efficient workflow and minimizes strain on staff. Consider factors such as equipment placement, working height, and accessibility to ensure a comfortable and productive work environment.

By taking these factors into account when selecting and maintaining equipment, you can create a foodservice establishment that prioritizes food safety, efficiency, and a comfortable working environment for staff members.

A foodservice facility's water supply and plumbing system are critical components in ensuring food safety and sanitation. The following are essential elements to consider when designing and maintaining these systems:
1. Potable water sources: Access to a reliable source of potable water is vital for food preparation, cleaning, and sanitizing. Ensure that the water supply meets local health regulations and is regularly tested for quality and safety.
2. Water temperature controls: Proper water temperature is necessary for effective cleaning, sanitizing, and food preparation. Install water heaters and mixing valves to maintain appropriate temperatures. For instance, hot water for dishwashing should typically be around 110-120°F (43-49°C) for washing and at least 180°F (82°C) for sanitizing in a high-temperature dishwasher.
3. Backflow prevention: To prevent the contamination of potable water, install backflow prevention devices in areas where there is a risk of cross-connection between potable and non-potable water, such as in connections to dishwashers, hose bibs, and chemical dispensers. These devices ensure that water flows in one direction only, safeguarding the clean water supply.
4. Proper drainage: Efficient drainage systems are necessary to eliminate wastewater and prevent standing water, which can attract pests and create unsanitary conditions. Install floor drains in areas where water is frequently used, such as kitchens and restrooms. Additionally, grease traps should be

installed to capture fats, oils, and grease from wastewater before it enters the sewer system, preventing blockages and backups.
5. Maintenance and monitoring: Regularly inspect and maintain the plumbing system to prevent leaks, corrosion, and other issues that could compromise the water supply's safety and quality. Monitor water pressure and flow rates, as inadequate pressure can affect the efficiency of cleaning and sanitizing equipment.

By incorporating these essential components into your foodservice facility's water supply and plumbing system, you can maintain a clean and sanitary environment that supports food safety and complies with relevant regulations.

Food Safety Regulations and Management Responsibilities

The Food Safety Regulations and Management Responsibilities chapter focuses on the legal requirements and best practices that foodservice establishments must follow to ensure food safety. It also emphasizes the roles and responsibilities of management in implementing and maintaining these standards. Key topics covered in this chapter include:

1. Food safety regulations: Familiarize yourself with the local, state, and federal regulations that govern food safety in your area. These rules cover aspects like food handling, storage, preparation, and service, as well as employee hygiene and training requirements.
2. Hazard Analysis and Critical Control Points (HACCP): Understand the principles of HACCP, a systematic approach to identifying, evaluating, and controlling food safety hazards throughout the production process. Implementing an effective HACCP plan helps prevent foodborne illnesses and ensures compliance with food safety regulations.
3. Management responsibilities: Recognize the essential role of management in establishing a food safety culture within the establishment. Management is responsible for developing food safety policies and procedures, providing employee training, conducting regular audits and inspections, and addressing non-compliance issues promptly.
4. Employee training and education: Ensure that all staff members receive proper training in food safety practices, personal hygiene, and their specific job duties. Ongoing training and reinforcement of these principles are crucial to maintaining a high standard of food safety.
5. Record keeping and documentation: Maintain accurate and up-to-date records of food safety practices, employee training, equipment maintenance, and other relevant activities. Proper documentation demonstrates your establishment's commitment to food safety and can help in case of inspections or audits by regulatory authorities.

By understanding and applying the principles covered in this chapter, foodservice managers can effectively uphold food safety standards, protect the health of their customers, and comply with applicable regulations.

Food safety regulations at local, state, and federal levels play a crucial role in ensuring the safety and well-being of customers and staff in foodservice establishments. These regulations establish standards and guidelines that help to prevent foodborne illnesses and maintain a high level of food safety. Key aspects covered by these regulations include:

1. Food handling: Regulations outline proper handling practices, such as washing hands before handling food, avoiding cross-contamination, and using gloves or utensils when appropriate. These guidelines help ensure that food remains safe and free from contaminants throughout the handling process.
2. Storage: Proper food storage is crucial for maintaining food safety. Regulations dictate the appropriate temperatures for storing various types of food, as well as the separation of raw and cooked products to prevent cross-contamination. Additionally, they address the proper rotation of stock to ensure freshness and minimize the risk of spoilage.
3. Preparation: Food safety regulations guide the safe preparation of food, including cooking temperatures, thawing methods, and cooling procedures. These guidelines ensure that harmful pathogens are eliminated or reduced to safe levels during the cooking process and that food is prepared in a manner that minimizes the risk of contamination.
4. Service: Regulations also cover food service practices, such as using clean and sanitized utensils, maintaining proper hot and cold holding temperatures, and discarding food that has been held at unsafe temperatures for too long. These practices help ensure that food served to customers is safe to consume.
5. Employee hygiene: Personal hygiene is a critical factor in food safety. Regulations require foodservice workers to maintain proper hygiene, such as frequent handwashing, wearing hair restraints, and not working when ill, to reduce the risk of contamination.
6. Training requirements: Food safety regulations often mandate that foodservice establishments provide appropriate training for their employees in safe food handling practices, sanitation, and other food safety-related topics. This training helps to create a culture of food safety within the establishment and ensures that staff members are aware of and adhere to necessary food safety practices.

By adhering to local, state, and federal food safety regulations, foodservice establishments can maintain a high standard of food safety, protect public health, and reduce the risk of foodborne illnesses.

Hazard Analysis and Critical Control Points (HACCP) is a systematic approach to food safety, aimed at identifying and controlling potential hazards throughout the food production process. By implementing an effective HACCP plan, foodservice establishments can prevent foodborne illnesses and ensure compliance with food safety regulations. The HACCP system is built around seven core principles:

1. Conduct a hazard analysis: This involves identifying all potential biological, chemical, and physical hazards associated with each step of the food production process. These hazards could cause foodborne illnesses if not properly controlled.
2. Identify critical control points (CCPs): CCPs are specific steps in the process where controls can be applied to prevent, eliminate, or reduce a hazard to an acceptable level. For example, cooking food to a specific temperature to kill pathogens is a critical control point.
3. Establish critical limits: These are the measurable criteria that must be met at each CCP to ensure hazards are effectively controlled. For instance, cooking poultry to an internal temperature of 165°F (74°C) for a specific time is a critical limit.
4. Set up monitoring procedures: Monitoring procedures involve regularly checking the CCPs to ensure critical limits are consistently met. This could involve taking temperature readings or visually inspecting for potential contaminants.
5. Establish corrective actions: If a critical limit is not met at a CCP, corrective actions must be taken to address the issue and prevent unsafe food from reaching customers. This could involve re-cooking, re-processing, or discarding the affected food.
6. Implement verification procedures: Verification involves assessing the HACCP plan to ensure its effectiveness in controlling hazards. This can include reviewing monitoring records, conducting internal audits, or validating equipment calibration.
7. Maintain documentation and records: Keeping accurate records of the HACCP plan, monitoring results, and any corrective actions taken is essential for demonstrating compliance with food safety regulations and facilitating continuous improvement of the food safety system.

By following these seven principles, foodservice establishments can effectively manage and control potential hazards in their operations. Implementing a robust HACCP plan not only prevents foodborne illnesses but also demonstrates a commitment to food safety and compliance with relevant regulations.

Establishing and maintaining a food safety culture within a foodservice establishment is crucial to ensuring the well-being of customers and staff. As part of their responsibility, foodservice management must focus on several key areas to create a strong food safety culture:

1. Development of policies and procedures: Management should create clear, comprehensive, and up-to-date policies and procedures that outline food safety expectations, practices, and standards. These documents should be easily accessible and understood by all employees.
2. Employee training: Providing ongoing and comprehensive training to staff is essential for fostering a food safety culture. This includes initial orientation, job-specific training, and regular refresher courses. Training should cover topics such as personal hygiene, proper food handling, allergen awareness, and hazard identification and control.
3. Regular audits and inspections: Management should schedule and conduct regular internal audits and inspections to assess the effectiveness of the food safety program. By identifying areas for improvement and addressing potential risks, management can continuously enhance the establishment's food safety practices.
4. Addressing non-compliance issues: When non-compliance issues are identified, management must take prompt corrective action. This may include additional training, disciplinary measures, or procedural changes. Addressing non-compliance sends a clear message that food safety is a priority and encourages a culture of accountability.
5. Leading by example: Management should actively demonstrate their commitment to food safety by adhering to the same policies and procedures as staff. This visible commitment helps reinforce the importance of food safety and sets a positive example for employees to follow.

By focusing on these key areas, foodservice management can establish and maintain a strong food safety culture within their establishment. This proactive approach not only protects customers and staff but also helps ensure compliance with food safety regulations and fosters a positive reputation for the business.

Employee training and education in food safety practices, personal hygiene, and job-specific duties are vital components of maintaining high standards of food safety within a foodservice establishment. Investing in ongoing reinforcement of these topics has multiple benefits for both the business and its customers.

1. Minimizing foodborne illnesses: Proper training helps employees understand the risks associated with improper food handling, preparation, and storage. When employees are well-versed in food safety practices, they can minimize the likelihood of foodborne illnesses, protecting customers and the establishment's reputation.
2. Compliance with regulations: Food safety regulations mandate that foodservice employees receive appropriate training to ensure safe food handling practices.

Ongoing training ensures that staff remains knowledgeable about current regulations and any updates that may occur.
3. Personal hygiene: Training on personal hygiene, such as handwashing, proper use of gloves, and appropriate attire, is essential for reducing the risk of cross-contamination and maintaining a clean work environment. Reinforcing these practices helps employees develop good habits that contribute to overall food safety.
4. Job-specific duties: Employees must understand their job-specific duties related to food safety, such as proper cooking temperatures, cleaning and sanitizing procedures, or allergen awareness. Ongoing training ensures that staff members are well-equipped to perform their tasks safely and effectively.
5. Continuous improvement: Regular reinforcement of food safety training helps employees stay engaged and motivated to maintain high standards. This fosters a culture of continuous improvement, where staff members are encouraged to seek opportunities for growth and learning.

By prioritizing employee training and education, foodservice establishments can create a strong food safety culture that promotes compliance with regulations, reduces the risk of foodborne illnesses, and enhances the overall quality of food and service.

Record-keeping and documentation are essential aspects of a foodservice establishment's commitment to food safety. Maintaining accurate and up-to-date records demonstrates to regulatory authorities that the establishment is proactive in ensuring compliance with food safety regulations. These records are invaluable during inspections or audits as they provide evidence of the establishment's adherence to proper procedures.

Some key types of records that should be maintained include:
1. Employee training records: These documents demonstrate that employees have received appropriate training in food safety, personal hygiene, and job-specific duties. They may include certificates, training session attendance logs, and assessments of employee competency.
2. HACCP records: As part of a Hazard Analysis and Critical Control Points (HACCP) plan, it is crucial to maintain records of monitoring, corrective actions, and verification activities. These records can help identify trends, validate that control measures are effective, and provide evidence of compliance during audits.
3. Temperature logs: Maintaining records of temperature checks for refrigeration units, freezers, and cooking equipment ensures that food is stored and prepared at the proper temperatures, reducing the risk of foodborne illnesses.

4. Cleaning and sanitizing schedules: Documentation of regular cleaning and sanitizing procedures helps verify that the establishment maintains a sanitary environment. This may include logs of when tasks were completed, the chemicals used, and any equipment maintenance performed.
5. Pest control records: Records of pest control measures, such as inspection dates, treatments applied, and any identified issues, demonstrate a proactive approach to maintaining a pest-free environment.
6. Supplier records: Documentation of supplier information, including product specifications, delivery records, and certificates of analysis, can help establish the safety and quality of ingredients used in the establishment.

By maintaining comprehensive records, foodservice establishments can demonstrate their commitment to food safety and ensure a higher level of preparedness for inspections or audits. Accurate record-keeping also helps identify areas for improvement, ultimately leading to a safer and more successful foodservice operation.

Practice Exam Section:

Welcome to the practice exam section! This part of the study guide has been designed to help you assess your understanding of the material and reinforce your learning. We believe that practice exams play a crucial role in preparing for the exam, allowing you to test your knowledge and get a feel for the types of questions you will encounter during the actual test. In fact, according to studies, one of the most important factors of success is practice questions. the more practice questions you take, the higher you score.

We have taken a unique approach in this section by providing the answer directly after each question. The rationale behind this decision is to prevent unnecessary page flipping, streamline your learning experience, and make it more efficient. By having the answers readily available, you can immediately review and understand the correct response without losing your train of thought or having to search through the guide. So make sure to hide the answer and prevent peaking by using paper or something of that manner.

This format encourages an interactive and engaging experience, allowing you to learn from any mistakes and reinforce your understanding of the concepts.

As you work through these practice questions, remember to stay focused, be patient with yourself, and use the answers as a learning tool to further solidify your knowledge. Good luck, and happy practicing!

1. Which of the following bacteria is most commonly associated with improper cooling and reheating of food, especially rice?
a. Salmonella
b. Escherichia coli
c. Bacillus cereus
d. Listeria monocytogenes

Answer: c. Bacillus cereus
Explanation: Bacillus cereus is commonly found in soil and can contaminate crops like rice. Improper cooling and reheating of food, particularly rice, can allow the bacteria to produce toxins that cause foodborne illness.

2. The danger zone, in which bacteria multiply rapidly, is the temperature range of:
a. 32°F to 60°F
b. 40°F to 140°F
c. 50°F to 150°F
d. 60°F to 160°F

Answer: b. 40°F to 140°F
Explanation: The danger zone refers to the temperature range of 40°F to 140°F, where bacteria can multiply rapidly, increasing the risk of foodborne illness. To minimize this risk, perishable foods should be kept out of this temperature range as much as possible.

3. Which microorganism can cause foodborne illness even if food is thoroughly cooked, due to the production of heat-stable toxins?
a. Campylobacter jejuni
b. Clostridium perfringens
c. Staphylococcus aureus
d. Vibrio cholerae

Answer: c. Staphylococcus aureus
Explanation: Staphylococcus aureus can produce heat-stable toxins that are not destroyed by normal cooking temperatures. Consuming food contaminated with these toxins can cause foodborne illness, even if the food has been thoroughly cooked.

4. Which of the following is a common foodborne parasite that can be transmitted through undercooked pork?
a. Giardia lamblia
b. Trichinella spiralis
c. Taenia solium
d. Cryptosporidium parvum

Answer: b. Trichinella spiralis
Explanation: Trichinella spiralis is a parasitic roundworm that can be transmitted to humans through the consumption of undercooked pork. Thorough cooking of pork to a minimum internal temperature of 145°F can help prevent infection.

5. What is the primary reason for using a sanitizer on food contact surfaces, in addition to thorough cleaning?
a. To remove visible dirt and debris
b. To neutralize odors
c. To reduce the number of microorganisms to safe levels
d. To dissolve grease and oils

Answer: c. To reduce the number of microorganisms to safe levels
Explanation: Sanitizers are used on food contact surfaces after cleaning to reduce the number of microorganisms to safe levels, minimizing the risk of foodborne illness. Sanitizers should be used at the proper concentration and contact time, as indicated by the manufacturer, for maximum effectiveness.

6. Which of the following foodborne illnesses is caused by a virus and often associated with the consumption of raw or undercooked shellfish?
a. E. coli infection
b. Salmonellosis
c. Norovirus
d. Botulism

Answer: c. Norovirus
Explanation: Norovirus is a viral foodborne illness that is commonly associated with the consumption of raw or undercooked shellfish, contaminated water, and contact with infected individuals. Proper food handling and hygiene practices can help reduce the risk of norovirus infection.

7. Which type of foodborne illness is commonly associated with undercooked poultry and can lead to Guillain-Barré Syndrome in severe cases?
a. Listeriosis
b. Campylobacteriosis
c. Cryptosporidiosis
d. Cyclosporiasis

Answer: b. Campylobacteriosis
Explanation: Campylobacteriosis is a bacterial foodborne illness caused by Campylobacter species. It is commonly associated with the consumption of undercooked poultry and can lead to severe complications like Guillain-Barré Syndrome in some cases.

8. Which foodborne illness is caused by a parasite and is often associated with the consumption of raw or undercooked meat, particularly pork and beef?
a. Toxoplasmosis
b. Giardiasis
c. Cyclosporiasis
d. Trichinellosis

Answer: a. Toxoplasmosis
Explanation: Toxoplasmosis is a parasitic foodborne illness caused by Toxoplasma gondii. It is often associated with the consumption of raw or undercooked meat, particularly pork and beef. Proper cooking and food handling practices can help prevent infection.

9. Which foodborne illness is caused by a fungus and can result from consuming moldy grains or nuts?
a. Aflatoxicosis
b. Anisakiasis
c. Listeriosis
d. Shigellosis

Answer: a. Aflatoxicosis
Explanation: Aflatoxicosis is a foodborne illness caused by the consumption of food contaminated with aflatoxins, which are toxic compounds produced by certain types of mold. Aflatoxins can contaminate grains and nuts, and proper storage and handling practices can help prevent contamination.

10. Which type of foodborne illness can be caused by the consumption of food contaminated with bacterial toxins, rather than the bacteria themselves?
a. Intoxication
b. Infection
c. Infestation
d. Ingestion

Answer: a. Intoxication
Explanation: Intoxication is a type of foodborne illness caused by the consumption of food contaminated with bacterial toxins, rather than the bacteria themselves. Examples of intoxication include illnesses caused by the toxins produced by Staphylococcus aureus and Clostridium botulinum.

11. Which of the following symptoms is commonly associated with foodborne illnesses caused by bacterial infections such as Salmonella and E. coli?
a. Double vision
b. Diarrhea
c. Tingling in extremities
d. Shortness of breath

Answer: b. Diarrhea
Explanation: Diarrhea is a common symptom of many foodborne illnesses caused by bacterial infections, such as Salmonella and E. coli. These infections often result from consuming contaminated food, and the symptoms can range from mild to severe.

12. What is a primary source of the foodborne illness Listeriosis, which can cause flu-like symptoms and even lead to life-threatening complications?
a. Raw eggs
b. Undercooked chicken
c. Deli meats and soft cheeses
d. Shellfish

Answer: c. Deli meats and soft cheeses. Explanation: Listeriosis is a foodborne illness caused by Listeria monocytogenes, which can be found in deli meats and soft cheeses. Pregnant women, newborns, older adults, and people with weakened immune systems are particularly at risk for severe complications from listeriosis.

13. Which foodborne illness is typically associated with the consumption of undercooked rice and can cause symptoms such as diarrhea, vomiting, and abdominal pain?
a. Bacillus cereus
b. Clostridium perfringens
c. Shigella
d. Campylobacter

Answer: a. Bacillus cereus. Explanation: Bacillus cereus is a foodborne illness commonly associated with the consumption of undercooked rice. The symptoms include diarrhea, vomiting, and abdominal pain, and proper cooking and storage practices can help prevent this illness.

14. Which foodborne illness can cause severe neurological symptoms, such as blurred vision, muscle weakness, and difficulty swallowing, and is commonly linked to improperly canned or preserved foods?
a. Salmonellosis
b. Botulism
c. Cryptosporidiosis
d. Cyclosporiasis

Answer: b. Botulism. Explanation: Botulism is a foodborne illness caused by the toxin produced by Clostridium botulinum bacteria. It can cause severe neurological symptoms and is often linked to improperly canned or preserved foods. Proper food preservation techniques can help prevent botulism.

15. Which foodborne illness is often associated with the consumption of raw or undercooked eggs, poultry, or meat, and can cause symptoms such as fever, abdominal cramps, and diarrhea?
a. Salmonellosis
b. Listeriosis
c. Giardiasis
d. Trichinellosis

Answer: a. Salmonellosis. Explanation: Salmonellosis is a foodborne illness caused by Salmonella bacteria. It is commonly associated with the consumption of raw or undercooked eggs, poultry, or meat. The symptoms include fever, abdominal cramps, and diarrhea, and proper food handling and cooking practices can help prevent infection.

16. Which of the following foodborne pathogens is known to produce heat-resistant spores and can cause symptoms such as diarrhea, vomiting, and abdominal pain?
a. Salmonella
b. Listeria monocytogenes
c. Bacillus cereus
d. Staphylococcus aureus

Answer: c. Bacillus cereus. Explanation: Bacillus cereus is a foodborne pathogen that produces heat-resistant spores. It can cause symptoms such as diarrhea, vomiting, and abdominal pain, and is commonly associated with the consumption of undercooked rice.

17. Which foodborne pathogen is able to survive and grow at refrigeration temperatures, making it a concern in ready-to-eat and deli foods?
a. Listeria monocytogenes
b. Campylobacter jejuni
c. Escherichia coli O157:H7
d. Vibrio cholerae

Answer: a. Listeria monocytogenes. Explanation: Listeria monocytogenes is a foodborne pathogen that can survive and grow at refrigeration temperatures. It is a concern in ready-to-eat and deli foods and can cause listeriosis, a severe illness that can lead to life-threatening complications in vulnerable individuals.

18. Which foodborne pathogen is most commonly associated with undercooked poultry and can cause campylobacteriosis, characterized by diarrhea, fever, and abdominal cramps?
a. Campylobacter jejuni
b. Clostridium botulinum
c. Salmonella
d. Shigella

Answer: a. Campylobacter jejuni
Explanation: Campylobacter jejuni is a foodborne pathogen commonly associated with undercooked poultry. It can cause campylobacteriosis, characterized by diarrhea, fever, and abdominal cramps. Proper cooking and handling of poultry can help prevent infection.

19. Which foodborne pathogen produces a potent toxin that can cause severe symptoms such as double vision, difficulty swallowing, and muscle weakness, and is often associated with improperly canned foods?
a. Clostridium perfringens
b. Clostridium botulinum
c. Staphylococcus aureus
d. Vibrio parahaemolyticus

Answer: b. Clostridium botulinum
Explanation: Clostridium botulinum is a foodborne pathogen that produces a potent toxin. It can cause severe symptoms such as double vision, difficulty swallowing, and muscle weakness, and is often associated with improperly canned foods. Proper food preservation techniques can help prevent botulism.

20. Which foodborne pathogen is often linked to the consumption of raw or undercooked beef, particularly ground beef, and can cause severe symptoms like bloody diarrhea and kidney failure in severe cases?
a. Shigella
b. Escherichia coli O157:H7
c. Salmonella
d. Norovirus

Answer: b. Escherichia coli O157:H7
Explanation: Escherichia coli O157:H7 is a foodborne pathogen often linked to the consumption of raw or undercooked beef, particularly ground beef. It can cause severe symptoms like bloody diarrhea and, in severe cases, kidney failure. Proper cooking and handling of beef can help prevent infection.

21. Which of the following is an example of a biological contaminant in food?
a. Glass shards
b. Pesticide residue
c. Bacteria
d. Cleaning chemicals

Answer: c. Bacteria
Explanation: Biological contaminants include microorganisms like bacteria, viruses, parasites, and fungi. These contaminants can cause foodborne illnesses if not properly controlled during food handling and preparation.

22. Which of the following is a common source of chemical contamination in food?
a. Hair
b. Metal fragments
c. Cleaning chemicals
d. Insects

Answer: c. Cleaning chemicals
Explanation: Chemical contamination can occur when chemicals such as cleaning agents, pesticides, or food additives are present in food. Proper storage and handling of chemicals and food items can help prevent chemical contamination.

23. What type of contaminant is most likely to be found in food due to improper handling of raw materials or damaged packaging?
a. Biological
b. Chemical
c. Physical
d. Radiological

Answer: c. Physical
Explanation: Physical contaminants are foreign objects, such as glass, metal, wood, or plastic, that can accidentally contaminate food during production, storage, or preparation. Proper handling of raw materials and packaging can help prevent physical contamination.

24. Which of the following is considered a chemical contaminant that can be introduced to food through the use of pesticides during agricultural production?
a. Salmonella
b. Glass fragments
c. Organophosphates
d. Insect parts

Answer: c. Organophosphates
Explanation: Organophosphates are a class of pesticides that can be introduced to food during agricultural production, making them a type of chemical contaminant. Proper pesticide application and adherence to regulations can help reduce the risk of contamination.

25. Which of the following scenarios is an example of physical contamination of food?
a. A food handler sneezes on a salad
b. A can of food is dented, allowing bacteria to enter
c. A broken lightbulb causes glass shards to fall into a container of flour
d. Pesticides are used on crops and remain as residue on vegetables

Answer: c. A broken lightbulb causes glass shards to fall into a container of flour
Explanation: Physical contamination occurs when foreign objects, like glass shards, are accidentally introduced to food. In this scenario, the broken lightbulb causing glass to fall into the flour is an example of physical contamination. Proper handling and inspection of food and equipment can help prevent such incidents.

26. What is the recommended method for thawing frozen food in order to prevent the growth of harmful microorganisms?
a. At room temperature on the counter
b. In a microwave on a low setting
c. In the refrigerator at 40°F (4°C) or below
d. In hot water

Answer: c. In the refrigerator at 40°F (4°C) or below. Explanation: Thawing frozen food in the refrigerator at 40°F (4°C) or below is the safest method because it keeps the food out of the temperature danger zone (41-135°F or 5-57°C), where bacteria can grow rapidly.

27. What is the most effective way to prevent cross-contamination between raw and ready-to-eat foods in a commercial kitchen?
a. Washing hands regularly
b. Using separate cutting boards and utensils
c. Cooking food to the proper internal temperature
d. Storing food at the correct temperature

Answer: b. Using separate cutting boards and utensils. Explanation: Using separate cutting boards, knives, and other utensils for raw and ready-to-eat foods can effectively prevent cross-contamination, which occurs when harmful microorganisms are transferred from one food item to another.

28. Which of the following practices can help prevent the growth of harmful microorganisms in food during storage?
a. Storing raw meat above ready-to-eat foods in the refrigerator
b. Storing food in uncovered containers
c. Maintaining cold holding temperatures at or below 41°F (5°C)
d. Storing food directly on the floor

Answer: c. Maintaining cold holding temperatures at or below 41°F (5°C)
Explanation: Keeping cold holding temperatures at or below 41°F (5°C) can help prevent the growth of harmful microorganisms in food during storage, as it keeps the food out of the temperature danger zone.

29. What is the most effective method for preventing the spread of foodborne illnesses caused by viruses in a foodservice establishment?
a. Cooking food to the proper internal temperature
b. Washing hands thoroughly and frequently
c. Freezing food to kill bacteria
d. Using a sanitizer solution on surfaces

Answer: b. Washing hands thoroughly and frequently
Explanation: Thorough and frequent handwashing is crucial in preventing the spread of foodborne illnesses caused by viruses, as these pathogens can be easily transferred from hands to food or surfaces if proper hygiene practices are not followed.

30. To ensure proper cooling and prevention of bacterial growth, what is the maximum time allowed for cooling cooked food from 135°F (57°C) to 41°F (5°C)?
a. 2 hours
b. 4 hours
c. 6 hours
d. 8 hours

Answer: b. 4 hours
Explanation: The maximum time allowed for cooling cooked food from 135°F (57°C) to 41°F (5°C) is 4 hours. This timeframe ensures that the food does not remain in the temperature danger zone, where harmful microorganisms can grow rapidly, for an extended period.

31. Which of the following is a critical aspect of personal hygiene for food handlers to prevent the spread of foodborne illnesses?
a. Wearing jewelry while handling food
b. Washing hands only after using the restroom
c. Wearing a hair restraint while handling food
d. Using a single-use glove instead of washing hands

Answer: c. Wearing a hair restraint while handling food
Explanation: Wearing a hair restraint, such as a hairnet or hat, while handling food is essential for food handlers to prevent hair, which can carry bacteria and other contaminants, from falling into food and causing contamination.

32. When should food handlers wash their hands to prevent the spread of foodborne illnesses?
a. After touching their face, hair, or clothing
b. Only before starting their shift
c. Only after handling raw meat
d. Only when visibly dirty

Answer: a. After touching their face, hair, or clothing
Explanation: Food handlers should wash their hands frequently, including after touching their face, hair, or clothing, to prevent the spread of foodborne illnesses. Hands can easily pick up harmful microorganisms from these surfaces, which can then be transferred to food if proper handwashing is not practiced.

33. What should a food handler do if they are experiencing symptoms of a foodborne illness, such as vomiting or diarrhea?
a. Wear gloves while handling food
b. Work only in areas where they are not handling food
c. Stay home or be reassigned to tasks that do not involve food handling
d. Continue working but inform their supervisor of their symptoms

Answer: c. Stay home or be reassigned to tasks that do not involve food handling
Explanation: Food handlers who are experiencing symptoms of a foodborne illness should stay home or be reassigned to tasks that do not involve food handling to prevent the spread of the illness to customers and coworkers.

34. Which of the following is a requirement for food handlers regarding fingernails?
a. Fingernails must be painted with a bright color
b. Fingernails must be kept short and clean
c. Fingernails must be covered with a bandage
d. Fingernail length does not matter

Answer: b. Fingernails must be kept short and clean
Explanation: Food handlers must keep their fingernails short and clean to prevent the spread of foodborne illnesses. Long or dirty fingernails can harbor harmful microorganisms, which can be transferred to food during handling.

35. Which of the following is NOT an acceptable method for food handlers to dry their hands after washing?
a. Using a single-use paper towel
b. Using a clean, dry cloth towel
c. Using an air dryer
d. Wiping hands on their apron

Answer: d. Wiping hands on their apron
Explanation: Wiping hands on an apron is not an acceptable method for drying hands after washing because aprons can be contaminated with harmful microorganisms. Food handlers should use single-use paper towels, clean cloth towels, or an air dryer to dry their hands to maintain proper hygiene and prevent contamination.

36. How does maintaining proper personal hygiene help prevent foodborne illnesses in foodservice establishments?
a. It reduces the risk of cross-contamination
b. It prevents the spread of harmful microorganisms from food handlers to food
c. It ensures that the food is cooked to the proper temperature
d. It makes the food taste better

Answer: b. It prevents the spread of harmful microorganisms from food handlers to food
Explanation: Maintaining proper personal hygiene is crucial because it helps prevent the spread of harmful microorganisms from food handlers to food. Good hygiene practices, such as frequent handwashing and wearing clean uniforms, minimize the risk of transmitting pathogens that can cause foodborne illnesses.

37. Which personal hygiene practice is most effective in reducing the risk of foodborne illnesses?
a. Washing hands for at least 20 seconds with soap and warm water
b. Washing hands with cold water only
c. Washing hands quickly with soap
d. Rinsing hands with water only

Answer: a. Washing hands for at least 20 seconds with soap and warm water
Explanation: Thoroughly washing hands for at least 20 seconds with soap and warm water is the most effective way to remove harmful microorganisms and prevent the spread of foodborne illnesses.

38. Why is it essential for food handlers to avoid touching their face, hair, or clothing while working with food?
a. It helps to maintain a professional appearance
b. It prevents the spread of pathogens from these surfaces to food
c. It ensures the correct cooking temperature is maintained
d. It makes the food more visually appealing

Answer: b. It prevents the spread of pathogens from these surfaces to food
Explanation: Food handlers should avoid touching their face, hair, or clothing while working with food because these surfaces can harbor harmful microorganisms. Touching these surfaces and then handling food can transfer pathogens to the food, increasing the risk of foodborne illnesses.

39. How can food handlers minimize the risk of spreading foodborne illnesses through contaminated clothing?
a. By wearing the same uniform for multiple shifts
b. By washing their uniforms with cold water only
c. By wearing clean uniforms each shift
d. By not wearing a uniform

Answer: c. By wearing clean uniforms each shift
Explanation: Wearing clean uniforms each shift helps minimize the risk of spreading foodborne illnesses through contaminated clothing. Dirty uniforms can harbor harmful microorganisms, which can be transferred to food during handling.

40. Why is it important for food handlers to cover cuts and wounds on their hands?
a. To prevent blood from getting on the food
b. To prevent the spread of pathogens from the wound to the food
c. To protect the wound from getting dirty
d. To make the food look more appetizing

Answer: b. To prevent the spread of pathogens from the wound to the food
Explanation: Covering cuts and wounds on the hands of food handlers is essential to prevent the spread of pathogens from the wound to the food. Open wounds can harbor harmful microorganisms, which can be transferred to food during handling, increasing the risk of foodborne illnesses.

41. What is the recommended minimum amount of time food handlers should spend washing their hands to effectively remove harmful microorganisms?
a. 5 seconds
b. 10 seconds
c. 20 seconds
d. 30 seconds

Answer: c. 20 seconds
Explanation: The recommended minimum amount of time food handlers should spend washing their hands is 20 seconds. This duration allows for the effective removal of harmful microorganisms, reducing the risk of foodborne illnesses.

42. Which of the following is an essential step in the handwashing process for food handlers?
a. Rinsing hands with cold water only
b. Scrubbing hands with soap for at least 20 seconds
c. Drying hands on a shared towel
d. Applying hand sanitizer without washing hands first

Answer: b. Scrubbing hands with soap for at least 20 seconds
Explanation: Scrubbing hands with soap for at least 20 seconds is a crucial step in the handwashing process for food handlers. This step ensures the effective removal of harmful microorganisms from the hands, reducing the risk of foodborne illnesses.

43. When should food handlers wash their hands?
a. Only at the beginning of their shift
b. After using the restroom, handling raw food, and touching their face or hair
c. Only when their hands are visibly dirty
d. Only after handling raw food

Answer: b. After using the restroom, handling raw food, and touching their face or hair
Explanation: Food handlers should wash their hands frequently, including after using the restroom, handling raw food, and touching their face or hair. Washing hands in these situations helps prevent the spread of harmful microorganisms that can cause foodborne illnesses.

44. Why is it important for food handlers to use warm water when washing their hands?
a. Warm water is more comfortable for the hands
b. Warm water is more effective in removing grease and dirt
c. Warm water makes the soap lather better
d. Both b and c

Answer: d. Both b and c
Explanation: Warm water is important for handwashing because it is more effective in removing grease and dirt and makes the soap lather better. These factors contribute to the effective removal of harmful microorganisms from the hands, reducing the risk of foodborne illnesses.

45. Which of the following is the recommended method for drying hands after washing?
a. Wiping hands on a shared towel
b. Shaking hands to remove excess water
c. Using a single-use paper towel or air dryer
d. Wiping hands on a personal apron

Answer: c. Using a single-use paper towel or air dryer
Explanation: The recommended method for drying hands after washing is to use a single-use paper towel or an air dryer. This ensures that hands remain clean and free of harmful microorganisms, reducing the risk of foodborne illnesses.

46. Which of the following is considered appropriate work attire for food handlers?
a. Loose-fitting clothing with long sleeves
b. A clean apron and close-fitting clothing
c. A watch and several bracelets
d. Open-toed shoes

Answer: b. A clean apron and close-fitting clothing
Explanation: Appropriate work attire for food handlers includes a clean apron and close-fitting clothing. This attire minimizes the risk of contamination from clothing and personal items, helping to ensure food safety.

47. Why is it important for food handlers to keep their fingernails short and clean?
a. Long nails can scratch coworkers
b. Long nails can harbor harmful microorganisms
c. Long nails are a fashion faux pas
d. Long nails make it difficult to grip utensils

Answer: b. Long nails can harbor harmful microorganisms
Explanation: Keeping fingernails short and clean is important for food handlers because long nails can harbor harmful microorganisms. Short, clean nails help reduce the risk of foodborne illnesses by minimizing the spread of pathogens.

48. What is the best practice for handling hair while working in a foodservice establishment?
a. Wearing hair down and loose
b. Wearing hair in a ponytail with a hairnet
c. Wearing a hat without a hairnet
d. Wearing a hairnet or hat to restrain hair

Answer: d. Wearing a hairnet or hat to restrain hair
Explanation: Wearing a hairnet or hat to restrain hair is the best practice for handling hair while working in a foodservice establishment. This practice helps prevent hair from contaminating food, reducing the risk of foodborne illnesses.

49. Which of the following items should be removed or covered while working with food?
a. Wedding ring
b. Plain band ring
c. Bracelet
d. Earrings

Answer: c. Bracelet
Explanation: Food handlers should remove or cover items such as bracelets while working with food. Jewelry, especially bracelets and watches, can harbor harmful microorganisms and pose a risk of contamination. Plain band rings and small earrings are generally acceptable, but it is important to follow the specific regulations and guidelines of the foodservice establishment.

50. How often should food handlers change their aprons?
a. Every hour
b. When they become soiled or contaminated
c. At the end of their shift
d. Once a week

Answer: b. When they become soiled or contaminated
Explanation: Food handlers should change their aprons when they become soiled or contaminated. Wearing a clean apron helps prevent cross-contamination and ensures a safe and hygienic food handling environment.

51. Which of the following symptoms should be reported to a supervisor immediately if experienced by a food handler?
a. Headache
b. Vomiting
c. Fatigue
d. Hunger

Answer: b. Vomiting
Explanation: Food handlers should report vomiting to a supervisor immediately, as it may be a sign of a foodborne illness. Reporting such symptoms ensures that sick employees are removed from food handling tasks, reducing the risk of contamination and potential illness for consumers.

52. A food handler with a sore throat and fever should:
a. Continue working, but avoid handling food directly
b. Stay home from work until symptoms improve
c. Work in a non-food handling position, such as dishwashing
d. Wear gloves while handling food

Answer: c. Work in a non-food handling position, such as dishwashing
Explanation: A food handler with a sore throat and fever should work in a non-food handling position, such as dishwashing, to minimize the risk of spreading illness. It is important to follow local regulations and guidelines regarding employee health in foodservice establishments.

53. If a food handler is diagnosed with a foodborne illness, they should:
a. Continue working and take medication
b. Be restricted from working with food until cleared by a medical professional
c. Work in a different area of the kitchen
d. Wear gloves and a mask while handling food

Answer: b. Be restricted from working with food until cleared by a medical professional
Explanation: A food handler diagnosed with a foodborne illness should be restricted from working with food until cleared by a medical professional. This practice helps prevent the spread of illness to consumers and maintains a safe and healthy food handling environment.

54. What is the main reason for reporting illnesses and managing sick food handlers?
a. To maintain a positive work environment
b. To prevent the spread of foodborne illnesses
c. To comply with labor laws
d. To ensure food handlers receive proper medical treatment

Answer: b. To prevent the spread of foodborne illnesses
Explanation: The main reason for reporting illnesses and managing sick food handlers is to prevent the spread of foodborne illnesses. By identifying and addressing potential health risks, foodservice establishments can maintain a safe environment and protect consumers from contaminated food.

55. Which action should a manager take when an employee reports symptoms of a foodborne illness?
a. Allow the employee to continue working with food
b. Require the employee to take a sick day
c. Assign the employee to a non-food handling task
d. Conduct an investigation to determine the cause of the illness

Answer: d. Conduct an investigation to determine the cause of the illness
Explanation: When an employee reports symptoms of a foodborne illness, a manager should conduct an investigation to determine the cause of the illness. The investigation will help identify potential risks and ensure appropriate actions are taken to protect consumers and maintain a safe food handling environment.

56. What is the primary purpose of an employee health policy in a foodservice establishment?
a. To ensure employees have access to healthcare benefits
b. To maintain a positive work environment
c. To minimize the risk of foodborne illness transmission
d. To comply with labor laws

Answer: c. To minimize the risk of foodborne illness transmission
Explanation: The primary purpose of an employee health policy is to minimize the risk of foodborne illness transmission. By implementing health policies and exclusion/restriction procedures, foodservice establishments can help prevent the spread of illness to consumers.

57. An employee health policy should include guidelines for:
a. Handling employee disputes and conflicts
b. Providing employee training and development
c. Reporting and managing symptoms of foodborne illnesses
d. Determining employee work schedules

Answer: c. Reporting and managing symptoms of foodborne illnesses
Explanation: An employee health policy should include guidelines for reporting and managing symptoms of foodborne illnesses. These guidelines help to ensure that sick food handlers are removed from food handling tasks, reducing the risk of contamination and potential illness for consumers.

58. Which of the following is an example of an exclusion procedure in a foodservice establishment?
a. Prohibiting employees with certain symptoms from working with food
b. Requiring employees to wear hair restraints
c. Restricting access to the kitchen for unauthorized personnel
d. Discarding expired food products

Answer: a. Prohibiting employees with certain symptoms from working with food
Explanation: Exclusion procedures are measures taken to prevent sick employees from working with food. This may include prohibiting employees with certain symptoms or diagnoses from food handling tasks, helping to reduce the risk of contamination and potential illness for consumers.

59. When should a food handler with a diagnosed foodborne illness be allowed to return to work?
a. After taking medication for 24 hours
b. When their symptoms have disappeared for at least 48 hours
c. When they have been cleared by a medical professional
d. After completing a food safety training course

Answer: c. When they have been cleared by a medical professional
Explanation: A food handler with a diagnosed foodborne illness should be allowed to return to work only when they have been cleared by a medical professional. This ensures that the risk of contamination and potential illness for consumers is minimized.

60. A food handler with a fever should:
a. Continue working, but avoid food handling tasks
b. Be sent home and not return until they are symptom-free
c. Be required to wear gloves while handling food
d. Be assigned to a non-food handling position, such as cleaning

Answer: d. Be assigned to a non-food handling position, such as cleaning
Explanation: A food handler with a fever should be assigned to a non-food handling position, such as cleaning, to minimize the risk of spreading illness. It is essential to follow local regulations and guidelines regarding employee health in foodservice establishments.

61. What is the primary reason for controlling time and temperature in TCS foods?
a. To maintain food quality and taste
b. To prevent the growth of harmful microorganisms
c. To reduce food waste
d. To comply with local regulations

Answer: b. To prevent the growth of harmful microorganisms
Explanation: Time and temperature control is crucial for TCS foods to prevent the growth of harmful microorganisms. TCS foods are more susceptible to bacterial growth, and controlling time and temperature helps ensure food safety.

62. The temperature danger zone, where harmful bacteria grow rapidly, is between:
a. 32°F and 140°F
b. 40°F and 140°F
c. 41°F and 135°F
d. 50°F and 120°F

Answer: c. 41°F and 135°F
Explanation: The temperature danger zone is between 41°F and 135°F. Within this range, harmful bacteria can grow rapidly, potentially causing foodborne illnesses.

63. When cooling TCS foods, they should be cooled from 135°F to 70°F within _____ hours and then to 41°F or lower within an additional _____ hours.
a. 2; 4
b. 4; 2
c. 2; 6
d. 6; 2

Answer: a. 2; 4
Explanation: When cooling TCS foods, they should be cooled from 135°F to 70°F within 2 hours and then to 41°F or lower within an additional 4 hours. Following these guidelines helps ensure that TCS foods do not remain in the temperature danger zone for an extended period, reducing the risk of bacterial growth.

64. When reheating TCS foods for hot holding, the foods should be reheated to a temperature of at least:
a. 135°F
b. 145°F
c. 155°F
d. 165°F

Answer: d. 165°F
Explanation: When reheating TCS foods for hot holding, the foods should be reheated to a temperature of at least 165°F for 15 seconds. This high temperature helps kill any harmful bacteria that may have grown during cooling or storage.

65. When holding TCS foods for cold storage, the maximum temperature they should be stored at is:
a. 32°F
b. 40°F
c. 41°F
d. 45°F

Answer: c. 41°F
Explanation: When holding TCS foods for cold storage, the maximum temperature they should be stored at is 41°F. Storing TCS foods at or below this temperature helps slow the growth of harmful bacteria and maintain food safety.

66. Which of the following is NOT considered a TCS food?
a. Sliced melons
b. Raw chicken
c. Hard-boiled eggs
d. Saltine crackers

Answer: d. Saltine crackers
Explanation: Time and Temperature Control for Safety (TCS) foods are those that require proper time and temperature control to prevent the growth of harmful microorganisms. Saltine crackers are not considered TCS foods because they are dry and low in moisture, which inhibits bacterial growth.

67. Which type of TCS food poses the highest risk for foodborne illness due to Salmonella?
a. Poultry
b. Shellfish
c. Fresh produce
d. Dairy products

Answer: a. Poultry
Explanation: Poultry poses the highest risk for foodborne illness due to Salmonella. Salmonella is commonly found in the intestinal tracts of poultry, and improper handling or undercooking can lead to contamination and foodborne illness.

68. Which TCS food category is most commonly associated with the foodborne illness Listeriosis?
a. Deli meats
b. Cooked vegetables
c. Fresh fruits
d. Seafood

Answer: a. Deli meats
Explanation: Listeriosis is most commonly associated with deli meats, as Listeria monocytogenes can grow at refrigeration temperatures and is often found in ready-to-eat foods like deli meats.

69. Why are raw or undercooked eggs considered high-risk TCS foods?
a. They can harbor Salmonella
b. They can contain E. coli
c. They can cause botulism
d. They can be contaminated with Staphylococcus aureus

Answer: a. They can harbor Salmonella
Explanation: Raw or undercooked eggs are considered high-risk TCS foods because they can harbor Salmonella. Salmonella is a common cause of foodborne illness, and proper cooking of eggs is necessary to kill the bacteria and prevent illness.

70. Which of the following TCS foods is most commonly associated with the foodborne illness Vibrio?
a. Ground beef
b. Shellfish
c. Poultry
d. Leafy greens

Answer: b. Shellfish
Explanation: Shellfish, particularly oysters, are most commonly associated with the foodborne illness Vibrio. Vibrio bacteria naturally occur in warm marine environments, and consuming raw or undercooked shellfish can lead to illness.

71. The temperature danger zone, where bacteria grow most rapidly, is between:
a. 20°F and 60°F
b. 32°F and 100°F
c. 41°F and 135°F
d. 50°F and 150°F

Answer: c. 41°F and 135°F
Explanation: The temperature danger zone is the range of temperatures in which bacteria can grow most rapidly, potentially leading to foodborne illness. This zone is between 41°F and 135°F.

72. Which of the following factors is NOT directly related to bacterial growth in food?
a. Time
b. Temperature
c. Moisture
d. Spiciness

Answer: d. Spiciness. Explanation: Spiciness does not directly affect bacterial growth in food. Time, temperature, and moisture are key factors that influence the growth of bacteria in food and must be properly controlled to ensure food safety.

73. The process of gradually heating or cooling food to keep it out of the temperature danger zone for as little time as possible is called:
a. Pasteurization
b. Thermalizing
c. Thawing
d. Tempering

Answer: d. Tempering. Explanation: Tempering is the process of gradually heating or cooling food to keep it out of the temperature danger zone for as little time as possible. This helps to minimize the growth of harmful microorganisms.

74. Which factor contributes to a food's acidity and can affect bacterial growth?
a. pH level
b. Oxygen content
c. Nutrient content
d. Temperature

Answer: a. pH level
Explanation: The pH level of a food contributes to its acidity, which can affect bacterial growth. Foods with a low pH (high acidity) inhibit the growth of bacteria, while foods with a high pH (low acidity) are more susceptible to bacterial growth.

75. Which of the following is NOT a recommended method for cooling hot foods to prevent bacterial growth?
a. Placing the food in shallow pans
b. Using an ice bath
c. Cooling the food at room temperature for several hours
d. Using a blast chiller

Answer: c. Cooling the food at room temperature for several hours
Explanation: Cooling food at room temperature for several hours is not a recommended method because it allows the food to remain in the temperature danger zone for too long, increasing the risk of bacterial growth. Using shallow pans, ice baths, or blast chillers are more effective methods for quickly cooling hot foods and minimizing bacterial growth.

76. What is the minimum internal cooking temperature for poultry, such as chicken or turkey?
a. 145°F
b. 155°F
c. 165°F
d. 175°F

Answer: c. 165°F
Explanation: The minimum internal cooking temperature for poultry, including chicken and turkey, is 165°F. Cooking poultry to this temperature ensures that harmful pathogens are destroyed, making the food safe to eat.

77. Which of the following methods is NOT recommended for cooling hot food quickly and safely?
a. Using a blast chiller
b. Placing the food in shallow pans
c. Leaving the food on the counter to cool for several hours
d. Submerging the container in an ice-water bath

Answer: c. Leaving the food on the counter to cool for several hours
Explanation: Leaving food on the counter to cool for several hours is not a recommended method, as it can allow the food to remain in the temperature danger zone for too long, increasing the risk of bacterial growth. Using a blast chiller, placing the food in shallow pans, or submerging the container in an ice-water bath are more effective methods for cooling hot food quickly and safely.

78. When reheating previously cooked food for hot holding, it must be heated to a minimum internal temperature of:
a. 135°F
b. 145°F
c. 155°F
d. 165°F

Answer: d. 165°F
Explanation: When reheating previously cooked food for hot holding, it must be heated to a minimum internal temperature of 165°F for at least 15 seconds. This ensures that any potential pathogens are destroyed and the food is safe to eat.

79. What is the proper method for thawing frozen food?
a. Leaving it on the counter at room temperature
b. Placing it in a container of hot water
c. Refrigerating it at a temperature of 41°F or lower
d. Microwaving it on high for 30 minutes

Answer: c. Refrigerating it at a temperature of 41°F or lower. Explanation: The proper method for thawing frozen food is refrigerating it at a temperature of 41°F or lower. This keeps the food out of the temperature danger zone and prevents the growth of harmful bacteria.

80. Which of the following is NOT a safe method for holding hot food at the proper temperature during service?
a. Using a steam table
b. Placing food in a preheated chafing dish
c. Leaving food in the oven at the lowest setting
d. Using a heated cabinet or hot holding unit

Answer: c. Leaving food in the oven at the lowest setting. Explanation: Leaving food in the oven at the lowest setting is not a safe method for holding hot food at the proper temperature during service, as it may not maintain the minimum required temperature of 135°F. Using a steam table, placing food in a preheated chafing dish, or using a heated cabinet or hot holding unit are safer methods for maintaining the correct temperature.

81. What is the minimum temperature for holding hot, ready-to-eat foods to minimize the risk of bacterial growth?
a. 115°F
b. 125°F
c. 135°F
d. 145°F

Answer: c. 135°F
Explanation: The minimum temperature for holding hot, ready-to-eat foods is 135°F. Maintaining this temperature helps to minimize the risk of bacterial growth and keeps food safe for consumption.

82. When using a cold buffet, potentially hazardous foods must be held at or below:
a. 32°F
b. 35°F
c. 41°F
d. 45°F

Answer: c. 41°F
Explanation: When using a cold buffet, potentially hazardous foods must be held at or below 41°F. This helps to prevent bacterial growth and ensures that the food remains safe for consumption.

83. What is the maximum time that cold food can be held without temperature control before it must be discarded?
a. 2 hours
b. 4 hours
c. 6 hours
d. 8 hours

Answer: b. 4 hours
Explanation: The maximum time that cold food can be held without temperature control is 4 hours. After this time, the food should be discarded to prevent the risk of bacterial growth and potential foodborne illness.

84. When using time as a public health control for hot food, what is the maximum time allowed before the food must be discarded?
a. 2 hours
b. 4 hours
c. 6 hours
d. 8 hours

Answer: b. 4 hours
Explanation: When using time as a public health control for hot food, the maximum time allowed before the food must be discarded is 4 hours. This helps to minimize the risk of bacterial growth and ensures that the food remains safe for consumption.

85. Which of the following is NOT a proper method for maintaining the temperature of hot food during transport?
a. Insulated food carriers
b. Heated cabinets
c. Placing the food in a cooler with ice packs
d. Electric hot holding units

Answer: c. Placing the food in a cooler with ice packs
Explanation: Placing hot food in a cooler with ice packs is not a proper method for maintaining the temperature of hot food during transport, as it would not maintain the required minimum temperature of 135°F. Insulated food carriers, heated cabinets, and electric hot holding units are suitable methods for maintaining the temperature of hot food during transport.

86. When using Time as a Public Health Control (TPHC), the food handler must do which of the following upon removing the food from temperature control?
a. Discard the food immediately
b. Serve the food within 30 minutes
c. Mark the food with the time it was removed from temperature control
d. Freeze the food for later use

Answer: c. Mark the food with the time it was removed from temperature control
Explanation: When using TPHC, the food handler must mark the food with the time it was removed from temperature control to ensure that it is served or discarded within the appropriate time frame.

87. Which of the following is a critical component of a written TPHC policy?
a. Reusing food that was held under TPHC
b. Monitoring food temperatures every hour
c. Discarding food that was held under TPHC after the maximum allowable time
d. Storing TPHC food with other food items

Answer: c. Discarding food that was held under TPHC after the maximum allowable time
Explanation: A critical component of a written TPHC policy is to discard food that was held under TPHC after the maximum allowable time. This helps to prevent the growth of harmful bacteria and ensures food safety.

88. When using TPHC, what is the maximum time that hot food can be held without temperature control before being served or discarded?
a. 2 hours
b. 4 hours
c. 6 hours
d. 8 hours

Answer: b. 4 hours
Explanation: The maximum time that hot food can be held without temperature control when using TPHC is 4 hours. After this time, the food must be served or discarded to prevent the risk of bacterial growth and potential foodborne illness.

89. Which of the following must be included in the documentation for TPHC?
a. The name and contact information of the food handler
b. The type of food and the time it was removed from temperature control
c. The temperature of the food when it was removed from temperature control
d. The expiration date of the food

Answer: b. The type of food and the time it was removed from temperature control
Explanation: The documentation for TPHC must include the type of food and the time it was removed from temperature control. This information helps to ensure proper tracking and compliance with TPHC guidelines.

90. What action should be taken if the food held under TPHC has exceeded the maximum allowable time?
a. The food should be reheated and served
b. The food should be frozen for later use
c. The food should be discarded
d. The food should be cooled and then reheated

Answer: c. The food should be discarded
Explanation: If the food held under TPHC has exceeded the maximum allowable time, it should be discarded. This helps to prevent the growth of harmful bacteria and ensures food safety.

91. What is the most effective way to prevent cross-contamination between raw and ready-to-eat foods in a commercial kitchen?
a. Store raw and ready-to-eat foods together in the same container
b. Use separate cutting boards and utensils for raw and ready-to-eat foods
c. Wash hands only at the beginning and end of food preparation
d. Cook all foods at the same temperature

Answer: b. Use separate cutting boards and utensils for raw and ready-to-eat foods
Explanation: Using separate cutting boards and utensils for raw and ready-to-eat foods is the most effective way to prevent cross-contamination. This practice helps to ensure that harmful bacteria from raw foods do not contaminate foods that will not be cooked before consumption.

92. Which of the following is NOT an effective allergen management practice in a commercial kitchen?
a. Washing hands and changing gloves between tasks
b. Using separate utensils and equipment for allergen-free foods
c. Placing allergen-free foods next to common allergen-containing foods
d. Clearly labeling allergen-free foods and ingredients

Answer: c. Placing allergen-free foods next to common allergen-containing foods
Explanation: Placing allergen-free foods next to common allergen-containing foods increases the risk of cross-contact, which can lead to allergic reactions for sensitive individuals. Effective allergen management practices involve separating allergen-free foods and ingredients from those containing common allergens.

93. Which of the following common allergens is often found in salad dressings and sauces?
a. Tree nuts
b. Soy
c. Shellfish
d. Wheat

Answer: b. Soy. Explanation: Soy is often found in salad dressings and sauces, as it can be an ingredient in soy sauce, mayonnaise, and other condiments. Being aware of common allergens in various food items helps to ensure proper allergen management in a commercial kitchen.

94. What is the most effective way to clean surfaces, utensils, and equipment to prevent allergen cross-contact?
a. Wipe with a dry cloth
b. Rinse with cold water
c. Wash with hot, soapy water, then rinse and sanitize
d. Use a sanitizing solution only

Answer: c. Wash with hot, soapy water, then rinse and sanitize. Explanation: Washing surfaces, utensils, and equipment with hot, soapy water, then rinsing and sanitizing, is the most effective way to prevent allergen cross-contact. This cleaning method helps to remove food particles and allergens from surfaces, reducing the risk of allergic reactions.

95. How should a food handler respond if a customer informs them of a food allergy?
a. Ignore the information, as it is not the food handler's responsibility
b. Inform the customer that the kitchen cannot accommodate food allergies
c. Communicate the allergy information to the chef or kitchen staff and follow proper allergen management procedures
d. Remove the allergen from the finished dish before serving

Answer: c. Communicate the allergy information to the chef or kitchen staff and follow proper allergen management procedures
Explanation: When a customer informs a food handler of a food allergy, the handler should communicate the allergy information to the chef or kitchen staff and follow proper allergen management procedures. This helps to ensure the safety of the customer and prevent allergic reactions.

96. Which of the following scenarios is most likely to cause cross-contamination in a commercial kitchen?
a. Washing hands frequently and thoroughly
b. Storing cleaning supplies separate from food items
c. Using the same cutting board for raw chicken and vegetables without cleaning and sanitizing it
d. Storing raw meat below ready-to-eat foods in the refrigerator

Answer: c. Using the same cutting board for raw chicken and vegetables without cleaning and sanitizing it
Explanation: Using the same cutting board for raw chicken and vegetables without cleaning and sanitizing it can cause cross-contamination, as harmful bacteria from the raw chicken can transfer to the vegetables, which might be consumed without further cooking.

97. When thawing frozen food, what method can help prevent cross-contamination?
a. Thawing food on the counter at room temperature
b. Thawing food in the refrigerator, separate from ready-to-eat foods
c. Thawing food in the microwave and then leaving it out at room temperature
d. Thawing food in a container of hot water

Answer: b. Thawing food in the refrigerator, separate from ready-to-eat foods
Explanation: Thawing food in the refrigerator, separate from ready-to-eat foods, helps prevent cross-contamination by keeping raw foods separate and at a safe temperature while thawing.

98. What should food handlers do with a dirty apron to prevent cross-contamination?
a. Continue wearing the dirty apron throughout their shift
b. Place the dirty apron on a food preparation surface
c. Hang the dirty apron on a hook in the kitchen
d. Place the dirty apron in a designated laundry area

Answer: d. Place the dirty apron in a designated laundry area
Explanation: Food handlers should place dirty aprons in a designated laundry area to prevent cross-contamination. This helps to keep dirty aprons away from food preparation surfaces and ready-to-eat foods.

99. What is the best practice to prevent cross-contamination when handling allergenic foods?
a. Washing hands and changing gloves before handling allergenic foods
b. Using the same utensils for allergenic and non-allergenic foods
c. Storing allergenic foods above non-allergenic foods in the refrigerator
d. Rinsing cutting boards used for allergenic foods with cold water

Answer: a. Washing hands and changing gloves before handling allergenic foods
Explanation: Washing hands and changing gloves before handling allergenic foods helps prevent cross-contamination and reduces the risk of causing an allergic reaction in sensitive individuals.

100. How should raw and cooked foods be stored in the refrigerator to minimize the risk of cross-contamination?
a. Store raw foods on the top shelf and cooked foods on the bottom shelf
b. Store raw and cooked foods on the same shelf
c. Store raw foods on the bottom shelf and cooked foods on the top shelf
d. Store raw and cooked foods in the same container

Answer: c. Store raw foods on the bottom shelf and cooked foods on the top shelf
Explanation: Storing raw foods on the bottom shelf and cooked foods on the top shelf of the refrigerator minimizes the risk of cross-contamination, as it prevents raw food juices from dripping onto cooked or ready-to-eat foods.

101. Case Study 1: Foodborne Illness Outbreak at a Wedding Reception

Background: A wedding reception was held at a local banquet hall with a buffet-style dinner service. Approximately 200 guests attended the event. The menu included a variety of dishes, such as grilled chicken, pasta salad, shrimp cocktail, and a chocolate fondue station for dessert. The food was prepared onsite by the venue's catering staff.

The problem: Within 24-48 hours after the wedding, numerous guests reported symptoms consistent with foodborne illness, including nausea, vomiting, diarrhea, and abdominal pain. The local health department was contacted to investigate the outbreak.

Question 1: What are the potential foodborne pathogens that could have caused this outbreak, and which foods might have been the source?

Question 2: What steps should the health department take to investigate the outbreak and identify the source of the contamination?

Question 3: What preventive measures could have been implemented by the catering staff to minimize the risk of foodborne illness at this event?

Case Study 2: Microbial Contamination in a Restaurant Kitchen

Background: A popular local restaurant has been in business for several years and has consistently received positive reviews. The menu offers a wide range of dishes, from steaks and seafood to salads and vegetarian options. The restaurant's owner takes pride in maintaining high standards for food safety and cleanliness.

The problem: During a routine health inspection, the health department discovered several instances of microbial contamination in the kitchen, including the presence of Listeria monocytogenes on a cutting board and Salmonella in a container of raw eggs. The health department issued a warning and demanded corrective actions.

Question 1: What are the potential consequences for the restaurant and its customers if the microbial contamination is not addressed?

Question 2: What steps should the restaurant owner take to address the microbial contamination and ensure the safety of the food served?

Question 3: What ongoing practices should be implemented in the restaurant to prevent future incidents of microbial contamination?

Case Study 1: Foodborne Illness Outbreak at a Wedding Reception

Answer 1: Potential foodborne pathogens that could have caused this outbreak include Salmonella, Campylobacter, Listeria, E. coli, Norovirus, and Staphylococcus aureus, among others. Foods that might have been the source include the shrimp cocktail, pasta salad, and grilled chicken, as these items are commonly associated with foodborne illnesses if not handled, stored, or cooked properly.

Answer 2: The health department should first interview affected individuals to gather information about the foods they consumed at the event and the onset of their symptoms. They should also collect samples of the leftover food and test them for the presence of common foodborne pathogens. Additionally, the health department should inspect the venue's kitchen and catering practices to identify any potential areas of concern, such as poor hygiene or inadequate temperature control.

Answer 3: Preventive measures that could have been implemented by the catering staff include proper handwashing, wearing gloves when handling ready-to-eat foods, ensuring adequate cooking and holding temperatures, avoiding cross-contamination between raw and cooked foods, and closely monitoring the temperature of foods during transportation and storage.

Case Study 2: Microbial Contamination in a Restaurant Kitchen

Answer 1: If the microbial contamination is not addressed, the restaurant's customers could be at risk of foodborne illness, which can cause severe symptoms and, in some cases, even death. The restaurant's reputation could also be severely damaged, leading to a decline in business, potential legal consequences, and possible closure.

Answer 2: The restaurant owner should take immediate action to address the microbial contamination. This includes proper cleaning and sanitizing of all contaminated surfaces and equipment, discarding any contaminated food, and retraining staff on proper food handling procedures. The owner should also implement a food safety management system, such as Hazard Analysis and Critical Control Points (HACCP), to identify potential hazards and establish control measures to prevent contamination.

Answer 3: Ongoing practices to prevent future incidents of microbial contamination include regular staff training on food safety and hygiene, strict adherence to food storage and temperature control guidelines, frequent cleaning and sanitizing of kitchen surfaces and equipment, and routinely monitoring and evaluating food safety practices. Additionally, the restaurant should establish a strong relationship with the local health department and work closely with them to ensure compliance with food safety regulations.

102. In a refrigerator, where should raw meats be stored to prevent cross-contamination?
a) On the top shelf
b) On a middle shelf
c) On the bottom shelf
d) Next to fruits and vegetables

Answer: c) On the bottom shelf. Explanation: Raw meats should be stored on the bottom shelf of a refrigerator to prevent cross-contamination. This way, any potential drips or leaks will not contaminate other food items stored below.

103. What is the ideal temperature range for dry storage of non-perishable food items?
a) 50-70°F (10-21°C)
b) 70-90°F (21-32°C)
c) 40-50°F (4-10°C)
d) 32-40°F (0-4°C)

Answer: a) 50-70°F (10-21°C). Explanation: The ideal temperature range for dry storage of non-perishable food items is 50-70°F (10-21°C). This helps to prevent the growth of bacteria and spoilage.

104. Which of the following storage techniques can help to prevent cross-contamination in a walk-in refrigerator?
a) Storing raw meats above ready-to-eat foods
b) Storing produce in tightly sealed containers
c) Placing all items directly on the floor
d) Mixing different food types on the same shelf

Answer: b) Storing produce in tightly sealed containers
Explanation: Storing produce in tightly sealed containers can help prevent cross-contamination in a walk-in refrigerator. This ensures that potential contaminants from other food items do not come into contact with the produce.

105. In a freezer, which of the following practices is most important for maintaining the quality and safety of frozen foods?
a) Regularly defrosting the freezer
b) Storing food in airtight, moisture-proof containers
c) Keeping the freezer door open for easy access
d) Stacking food items as high as possible

Answer: b) Storing food in airtight, moisture-proof containers
Explanation: Storing food in airtight, moisture-proof containers helps maintain the quality and safety of frozen foods. This practice prevents freezer burn and the growth of ice crystals, which can affect the texture and flavor of the food.

106. According to the food storage hierarchy, which of these food items should be stored on the highest shelf in the refrigerator?
a) Raw poultry
b) Raw ground meat
c) Cooked seafood
d) Whole fruits and vegetables

Answer: d) Whole fruits and vegetables
Explanation: Whole fruits and vegetables should be stored on the highest shelf in the refrigerator, according to the food storage hierarchy. This minimizes the risk of cross-contamination, as raw meats and other potentially hazardous items are stored below.

107. When should a cutting board be replaced in a commercial kitchen?
a) When it has deep grooves or cuts
b) After every use
c) Once a year
d) When it no longer matches the kitchen decor

Answer: a) When it has deep grooves or cuts
Explanation: A cutting board should be replaced in a commercial kitchen when it has deep grooves or cuts. These grooves can harbor harmful bacteria and make it difficult to properly clean and sanitize the cutting board.

108. Which of the following is the most effective method for sanitizing a cutting board after use?
a) Wiping it with a damp cloth
b) Rinsing it under hot water
c) Washing it with soap and water, then using a sanitizing solution
d) Soaking it in bleach for 10 minutes

Answer: c) Washing it with soap and water, then using a sanitizing solution
Explanation: The most effective method for sanitizing a cutting board after use is to wash it with soap and water, then use a sanitizing solution. This process ensures that any harmful bacteria or pathogens are effectively eliminated.

109. How should utensils be stored after cleaning and sanitizing to prevent contamination?
a) Stacked together in a drawer
b) In a container with the handles facing up
c) In a container with the handles facing down
d) Laid flat on a clean surface

Answer: c) In a container with the handles facing down
Explanation: Storing utensils in a container with the handles facing down helps prevent contamination. This way, when someone grabs a utensil, they only touch the handle, not the part that comes into contact with food.

110. What is the recommended water temperature for washing dishes in a commercial dishwasher?
a) 110°F (43°C)
b) 140°F (60°C)
c) 165°F (74°C)
d) 180°F (82°C)

Answer: d) 180°F (82°C)
Explanation: The recommended water temperature for washing dishes in a commercial dishwasher is 180°F (82°C). This high temperature ensures that harmful bacteria and pathogens are effectively eliminated during the wash cycle.

111. How often should a commercial kitchen's dishwasher be cleaned and sanitized?
a) After each use
b) At the end of each day
c) Weekly
d) Monthly

Answer: b) At the end of each day
Explanation: A commercial kitchen's dishwasher should be cleaned and sanitized at the end of each day. Daily cleaning and sanitizing help maintain the effectiveness of the dishwasher and prevent the buildup of bacteria or pathogens.

112. Which of the following is NOT considered one of the "Big 8" allergens?
a) Shellfish
b) Peanuts
c) Quinoa
d) Soy

Answer: c) Quinoa
Explanation: The "Big 8" allergens are milk, eggs, fish, crustacean shellfish, tree nuts, peanuts, wheat, and soy. Quinoa is not considered one of the "Big 8" allergens.

113. What is the best practice for preventing allergen cross-contact in a commercial kitchen?
a) Washing hands frequently
b) Using separate cutting boards and utensils for allergen-free food preparation
c) Cooking allergen-free foods first
d) All of the above

Answer: d) All of the above. Explanation: To prevent allergen cross-contact, it is essential to practice proper handwashing, use separate cutting boards and utensils for allergen-free food preparation, and cook allergen-free foods first.

114. How should a foodservice establishment handle a customer's request for an allergen-free meal?
a) Suggest that the customer orders a salad
b) Communicate the request to the chef and kitchen staff to ensure proper precautions are taken
c) Offer to remove the allergen from the dish after it has been prepared
d) Tell the customer that the kitchen cannot guarantee an allergen-free meal

Answer: b) Communicate the request to the chef and kitchen staff to ensure proper precautions are taken. Explanation: When a customer requests an allergen-free meal, it is crucial to communicate this request to the chef and kitchen staff to ensure that proper precautions are taken to prevent cross-contact and provide a safe meal for the customer.

115. Which allergen can be found in Worcestershire sauce?
a) Tree nuts
b) Fish
c) Soy
d) Wheat

Answer: b) Fish
Explanation: Worcestershire sauce contains anchovies, a type of fish, which is one of the "Big 8" allergens.

116. What action should be taken if a customer has an allergic reaction in a foodservice establishment?
a) Offer them a glass of water to help soothe their throat
b) Ask the customer to leave the establishment to prevent other customers from panicking
c) Call emergency services and inform them of the allergic reaction
d) Have the customer take an antihistamine to reduce symptoms

Answer: c) Call emergency services and inform them of the allergic reaction. Explanation: If a customer has an allergic reaction in a foodservice establishment, it is essential to call emergency services and inform them of the allergic reaction. An allergic reaction can be life-threatening, and prompt medical attention is crucial.

117. Which of the following is the most effective method for preventing allergen cross-contact when handling special orders in a commercial kitchen?
a) Designating a specific area for allergen-free meal preparation
b) Wiping down shared utensils with a damp cloth
c) Rinsing cutting boards with hot water
d) Using the same cutting board but preparing allergen-free meals first

Answer: a) Designating a specific area for allergen-free meal preparation. Explanation: Designating a specific area for allergen-free meal preparation helps prevent cross-contact and is the most effective method for ensuring the safety of special orders.

118. What is the best practice for communicating allergen information to kitchen staff when handling special orders?
a) Verbally informing the chef
b) Using a special ticket or order notation
c) Writing a note on the order and placing it with the other orders
d) Relying on the server to remember the allergen information

Answer: b) Using a special ticket or order notation. Explanation: Using a special ticket or order notation ensures that allergen information is clearly communicated to the kitchen staff and reduces the risk of mistakes when handling special orders.

119. When preparing an allergen-free meal, which of the following ingredients should be avoided?
a) Seasoning blends that may contain allergens
b) Fresh fruits and vegetables
c) Unprocessed meat and poultry
d) Oil that has not been used to cook other foods

Answer: a) Seasoning blends that may contain allergens
Explanation: Seasoning blends may contain allergens, such as wheat or soy, which can cause an allergic reaction. It's essential to read labels and avoid using ingredients that may contain allergens when preparing allergen-free meals.

120. In addition to using separate equipment for allergen-free meal preparation, what other step should be taken to ensure cross-contact is minimized?
a) Washing hands frequently and thoroughly
b) Using the same utensils for all ingredients
c) Cooking allergen-free meals at a lower temperature
d) Placing allergen-free meals on the same tray as other meals

Answer: a) Washing hands frequently and thoroughly
Explanation: Frequent and thorough handwashing helps prevent cross-contact and is a critical step in ensuring the safety of allergen-free meals.

121. Which of the following actions should be taken if a customer's special order is accidentally exposed to an allergen during preparation?
a) Serve the meal and hope the customer does not have a reaction
b) Attempt to remove the allergen from the dish before serving
c) Inform the customer of the error and start preparing a new meal
d) Substitute the dish with another allergen-free option without informing the customer

Answer: c) Inform the customer of the error and start preparing a new meal
Explanation: If a customer's special order is accidentally exposed to an allergen during preparation, it is crucial to inform the customer of the error and start preparing a new meal. This ensures the customer's safety and maintains transparency and trust.

122. Which of the following is the correct order of steps to properly clean and sanitize a surface in a foodservice establishment?
a) Rinse, sanitize, wash
b) Wash, rinse, sanitize
c) Sanitize, rinse, wash
d) Wash, sanitize, rinse

Answer: b) Wash, rinse, sanitize
Explanation: The correct order of steps to clean and sanitize a surface is to wash with detergent, rinse with water, and then apply a sanitizer. This process ensures that surfaces are free from dirt and debris before sanitizing.

123. What is the primary purpose of a pest management program in a foodservice facility?
a) To exterminate pests once an infestation is discovered
b) To prevent pest infestations through proactive measures
c) To maintain a certain level of acceptable pests
d) To rely on chemical pesticides as the primary method of control

Answer: b) To prevent pest infestations through proactive measures
Explanation: The primary purpose of a pest management program is to prevent pest infestations through proactive measures, such as proper sanitation, exclusion, and monitoring practices.

124. What is an essential component of an integrated pest management (IPM) program?
a) Frequent use of chemical pesticides
b) Ignoring the presence of pests until they become a problem
c) Regular inspections and monitoring for signs of pests
d) Storing food in open containers to identify potential pest problems

Answer: c) Regular inspections and monitoring for signs of pests. Explanation: Regular inspections and monitoring for signs of pests are essential components of an IPM program, as they help to identify potential issues early and allow for targeted, effective control measures.

125. Which of the following practices is the most effective way to prevent pests from entering a foodservice facility?
a) Leaving doors and windows open for ventilation
b) Using air curtains and screens on doors and windows
c) Placing traps near entrances and exits
d) Regularly applying pesticide sprays around the perimeter of the building

Answer: b) Using air curtains and screens on doors and windows. Explanation: Using air curtains and screens on doors and windows is the most effective way to prevent pests from entering a foodservice facility, as they create a barrier that deters pests without the use of chemicals.

126. What is the primary difference between cleaning and sanitizing in a foodservice environment?
a) Cleaning removes dirt and debris, while sanitizing kills bacteria and other microorganisms
b) Cleaning kills bacteria, while sanitizing removes dirt and debris
c) Cleaning is a daily task, while sanitizing is performed weekly
d) Cleaning requires the use of water, while sanitizing does not

Answer: a) Cleaning removes dirt and debris, while sanitizing kills bacteria and other microorganisms
Explanation: The primary difference between cleaning and sanitizing is that cleaning removes dirt, grease, and debris from surfaces, while sanitizing reduces the number of bacteria and other microorganisms to a safe level.

127. Which of the following best describes the primary function of cleaning agents in a foodservice environment?
a) To kill pathogens and reduce the risk of foodborne illness
b) To remove visible dirt, grease, and debris from surfaces
c) To maintain a pleasant aroma in the establishment
d) To create a shiny, polished appearance on surfaces

Answer: b) To remove visible dirt, grease, and debris from surfaces
Explanation: The primary function of cleaning agents is to remove visible dirt, grease, and debris from surfaces. Cleaning agents do not necessarily kill pathogens or sanitize surfaces.

128. Which of the following factors is most important when choosing a sanitizing agent for use in a foodservice establishment?
a) The aroma of the sanitizing agent
b) The ability of the agent to break down grease and dirt
c) The effectiveness of the agent against specific pathogens
d) The color and appearance of the sanitizing agent

Answer: c) The effectiveness of the agent against specific pathogens
Explanation: The most important factor when choosing a sanitizing agent is its effectiveness against specific pathogens that may be present in the foodservice environment. Sanitizing agents should be chosen based on their ability to kill or reduce the number of harmful microorganisms.

129. When using a two-compartment sink for cleaning and sanitizing, which of the following is the correct order of operations?
a) Wash in the first compartment, sanitize in the second compartment
b) Sanitize in the first compartment, wash in the second compartment
c) Rinse in the first compartment, wash in the second compartment
d) Wash in the first compartment, rinse in the second compartment

Answer: a) Wash in the first compartment, sanitize in the second compartment
Explanation: In a two-compartment sink, the correct order of operations is to wash items in the first compartment with detergent and water, and then sanitize in the second compartment using a sanitizing solution.

130. What is the primary purpose of using test strips in a foodservice environment?
a) To measure the pH level of cleaning solutions
b) To check the concentration of sanitizing solutions
c) To determine the presence of allergens on surfaces
d) To identify the type of pathogens present on surfaces

Answer: b) To check the concentration of sanitizing solutions
Explanation: Test strips are used to check the concentration of sanitizing solutions, ensuring that they are effective in killing or reducing the number of harmful microorganisms on surfaces.

131. Which of the following statements best describes the role of heat in the sanitizing process?
a) Heat helps to remove grease and debris from surfaces
b) Heat is only effective for sanitizing when used with chemical sanitizers
c) Heat alone can effectively sanitize surfaces and utensils
d) Heat is not an effective method of sanitizing surfaces

Answer: c) Heat alone can effectively sanitize surfaces and utensils
Explanation: Heat alone can effectively sanitize surfaces and utensils, as it can kill or reduce the number of harmful microorganisms when used at the appropriate temperature and duration.

132. Which of the following sanitizers is most effective against a broad range of microorganisms and is stable over a wide range of temperatures and pH levels?
a) Quaternary ammonium compounds (QUATs)
b) Iodine
c) Chlorine
d) Hydrogen peroxide

Answer: a) Quaternary ammonium compounds (QUATs)
Explanation: QUATs are effective against a broad range of microorganisms and are stable over a wide range of temperatures and pH levels. This makes them a versatile choice for sanitizing in various foodservice settings.

133. When using chlorine as a sanitizer, which of the following factors must be considered to ensure its effectiveness?
a) Concentration, contact time, and temperature
b) Color, aroma, and texture
c) Solubility, hardness, and pH level
d) Viscosity, density, and boiling point

Answer: a) Concentration, contact time, and temperature
Explanation: When using chlorine as a sanitizer, it is important to consider the concentration of the solution, the contact time with surfaces, and the temperature of the solution to ensure its effectiveness in killing microorganisms.

134. Iodine is an effective sanitizer when used at the correct concentration. However, it has some drawbacks, including:
a) Its inability to kill bacteria
b) Its corrosiveness to certain metals
c) Its tendency to discolor certain surfaces
d) Its ineffectiveness in the presence of organic material

Answer: c) Its tendency to discolor certain surfaces
Explanation: Iodine is an effective sanitizer, but it can discolor certain surfaces, which may be a concern in some foodservice settings.

135. Which of the following sanitizers is commonly used in foodservice settings due to its effectiveness at low concentrations and rapid action, but may be corrosive to some surfaces?
a) Hydrogen peroxide
b) Chlorine
c) Quaternary ammonium compounds (QUATs)
d) Iodine

Answer: b) Chlorine
Explanation: Chlorine is commonly used in foodservice settings because it is effective at low concentrations and acts rapidly. However, it can be corrosive to some surfaces, which is a factor to consider when choosing a sanitizer.

136. In a foodservice environment, which of the following is the most appropriate use for hydrogen peroxide as a sanitizer?
a) Surface sanitizing of cutting boards and countertops
b) Sanitizing utensils and dishware
c) Cleaning and sanitizing floors and walls
d) Hand sanitizing for food handlers

Answer: a) Surface sanitizing of cutting boards and countertops
Explanation: Hydrogen peroxide is an effective sanitizer for surface sanitizing of cutting boards and countertops. While it can be used for other purposes, it is most appropriate and commonly used for surface sanitizing in foodservice settings.

137. Which of the following steps is NOT a part of the correct procedure for cleaning and sanitizing utensils and equipment using a three-compartment sink?
a) Rinse with hot water in the first compartment
b) Wash with detergent in the second compartment
c) Rinse with clean water in the third compartment
d) Sanitize in the second compartment

Answer: d) Sanitize in the second compartment. Explanation: The correct procedure for cleaning and sanitizing utensils and equipment using a three-compartment sink involves rinsing in the first compartment, washing with detergent in the second compartment, and rinsing with clean water in the third compartment. Sanitizing should be done in a separate compartment or container after these steps.

138. In a foodservice operation, which of the following surfaces should be cleaned and sanitized after each use?
a) Floors
b) Walls
c) Cutting boards
d) Ceiling

Answer: c) Cutting boards. Explanation: Cutting boards should be cleaned and sanitized after each use to prevent cross-contamination and the growth of harmful microorganisms. Floors, walls, and ceilings should be cleaned regularly, but not necessarily after each use.

139. Which of the following factors is NOT important to consider when choosing a sanitizer for use in a foodservice operation?
a) The sanitizer's effectiveness against microorganisms
b) The sanitizer's effect on the appearance of surfaces
c) The price of the sanitizer
d) The flavor of the sanitizer

Answer: d) The flavor of the sanitizer. Explanation: The flavor of the sanitizer is not an important factor to consider when choosing a sanitizer for use in a foodservice operation. The effectiveness against microorganisms, the effect on the appearance of surfaces, and the price are more relevant factors to consider.

140. What is the correct order of steps for cleaning and sanitizing a cutting board in a foodservice operation?
a) Rinse, sanitize, wash with detergent, air dry
b) Wash with detergent, rinse, sanitize, air dry
c) Sanitize, rinse, wash with detergent, air dry
d) Air dry, sanitize, wash with detergent, rinse

Answer: b) Wash with detergent, rinse, sanitize, air dry
Explanation: The correct order of steps for cleaning and sanitizing a cutting board in a foodservice operation is to wash with detergent, rinse with clean water, sanitize, and then allow the cutting board to air dry.

141. Which of the following best describes the purpose of air drying utensils and equipment after cleaning and sanitizing in a foodservice operation?
a) To prevent cross-contamination from towels
b) To improve the appearance of the utensils and equipment
c) To save time and labor
d) To ensure the sanitizer remains effective

Answer: a) To prevent cross-contamination from towels
Explanation: Air drying utensils and equipment after cleaning and sanitizing helps prevent cross-contamination that could occur if towels were used. This is the primary reason for air drying in a foodservice operation.

142. Which of the following is a common sign of a rodent infestation in a foodservice facility?
a) Grease marks along walls
b) Chewed packaging materials
c) Small, irregularly shaped holes in walls
d) All of the above

Answer: d) All of the above
Explanation: Grease marks along walls, chewed packaging materials, and small, irregularly shaped holes in walls are all common signs of a rodent infestation in a foodservice facility.

143. What is an Integrated Pest Management (IPM) approach?
a) A method that relies solely on chemical pesticides to eliminate pests
b) A combination of pest control methods that focuses on prevention and control with minimal environmental impact
c) A system of controlling pests through regular cleaning and maintenance of facilities
d) A government-mandated pest control program for foodservice operations

Answer: b) A combination of pest control methods that focuses on prevention and control with minimal environmental impact. Explanation: Integrated Pest Management (IPM) is an approach that combines various pest control methods, emphasizing prevention and control with minimal environmental impact. It does not rely solely on chemical pesticides and is not a government-mandated program.

144. Which of the following is NOT a recommended pest prevention method for a foodservice operation?
a) Storing food in airtight containers
b) Leaving doors and windows open to improve air circulation
c) Regularly cleaning and sanitizing surfaces
d) Sealing gaps and cracks in the facility

Answer: b) Leaving doors and windows open to improve air circulation. Explanation: Leaving doors and windows open can actually increase the risk of pest entry. Instead, proper ventilation and air circulation should be maintained without leaving openings for pests. The other options are recommended pest prevention methods.

145. What is the primary reason for maintaining proper sanitation practices in a foodservice operation to prevent pests?
a) To eliminate potential breeding sites for pests
b) To comply with government regulations
c) To maintain a positive image for customers
d) To reduce the need for chemical pest control

Answer: a) To eliminate potential breeding sites for pests. Explanation: Maintaining proper sanitation practices in a foodservice operation helps eliminate potential breeding sites for pests, reducing the risk of infestation. While compliance with government regulations, maintaining a positive image, and reducing the need for chemical pest control are additional benefits, the primary reason is to eliminate breeding sites.

146. Which of the following is a common characteristic of cockroaches that makes them difficult to control in a foodservice environment?
a) They are attracted to light
b) They are primarily active during the day
c) They are resistant to most chemical pesticides
d) They can survive without food for long periods

Answer: d) They can survive without food for long periods. Explanation: Cockroaches can survive without food for long periods, making them difficult to control in a foodservice environment. Regular cleaning and sanitation practices, proper food storage, and other prevention measures are essential to minimize the risk of infestation.

147. What is the first step a foodservice establishment should take when they discover a pest infestation?
a) Immediately close the facility
b) Notify the local health department
c) Attempt to resolve the issue in-house
d) Contact a licensed pest control professional

Answer: d) Contact a licensed pest control professional
Explanation: When a pest infestation is discovered in a foodservice establishment, the first step should be to contact a licensed pest control professional. They will help assess the situation, determine the appropriate course of action, and ensure the issue is resolved safely and effectively.

148. Which of the following actions should be taken if live pests are found in or near food during an inspection?
a) Remove and discard the affected food items
b) Continue to use the food but increase sanitation efforts
c) Wait for the pests to leave the area before continuing operations
d) Use a pesticide to kill the pests and continue using the food

Answer: a) Remove and discard the affected food items. Explanation: If live pests are found in or near food during an inspection, the affected food items should be removed and discarded immediately. Using the food, waiting for the pests to leave, or using a pesticide would not be appropriate responses, as they could potentially compromise food safety.

149. What is the most appropriate action to take when you find evidence of pests in a dry storage area?
a) Ignore the issue and hope it resolves itself
b) Clean the area and inspect all stored products for contamination
c) Apply a pesticide to the area without consulting a professional
d) Move the products to another storage area and continue operations

Answer: b) Clean the area and inspect all stored products for contamination. Explanation: When evidence of pests is found in a dry storage area, the most appropriate action is to clean the area and inspect all stored products for contamination. Ignoring the issue or applying a pesticide without consulting a professional could lead to further problems, and moving the products without addressing the issue could spread the infestation.

150. What should be done with any contaminated food discovered during a pest infestation?
a) Cook the food to kill any pests or bacteria present
b) Donate the food to a local food bank
c) Dispose of the contaminated food properly
d) Sell the food at a discounted price

Answer: c) Dispose of the contaminated food properly
Explanation: Any contaminated food discovered during a pest infestation should be disposed of properly. Cooking the food, donating it, or selling it at a discounted price would not be appropriate, as it could pose a risk to consumers' health.

151. Which of the following is an important step to take after a pest infestation has been resolved?
a) Implement preventive measures to avoid future infestations
b) Reduce cleaning and sanitation efforts since the problem has been resolved
c) Reopen the facility without informing the local health department
d) Offer discounts to customers to rebuild the establishment's reputation

Answer: a) Implement preventive measures to avoid future infestations
Explanation: After a pest infestation has been resolved, it is important to implement preventive measures to avoid future infestations. This includes maintaining regular cleaning and sanitation efforts, addressing any structural issues, and monitoring for any signs of reinfestation. Reducing cleaning efforts, reopening without informing the health department, or offering discounts are not the primary concerns after resolving an infestation.

152. What is the primary purpose of a properly designed foodservice facility layout?
a) To increase the facility's aesthetic appeal
b) To maximize profitability by fitting in as many customers as possible
c) To provide a comfortable work environment for employees
d) To promote food safety and efficient workflow

Answer: d) To promote food safety and efficient workflow
Explanation: The primary purpose of a properly designed foodservice facility layout is to promote food safety and efficient workflow. While aesthetics, profitability, and employee comfort are important, they should not compromise the primary goal of ensuring food safety and efficiency.

153. Which of the following should be considered when designing a food preparation area?
a) Separating raw and ready-to-eat food preparation areas
b) Placing all equipment and workstations in a single area for easy access
c) Focusing on the visual appeal of the area rather than its functionality
d) Allocating the majority of the space for customer seating

Answer: a) Separating raw and ready-to-eat food preparation areas
Explanation: When designing a food preparation area, it is crucial to separate raw and ready-to-eat food preparation areas to prevent cross-contamination. Focusing on visual appeal, placing all equipment in one area, or prioritizing customer seating over food safety would not be appropriate choices.

154. What is an essential consideration when selecting flooring for a foodservice facility?
a) The color and design of the flooring
b) The ease of maintenance and cleaning
c) The cost of the flooring material
d) The type of flooring used in other businesses in the area

Answer: b) The ease of maintenance and cleaning
Explanation: When selecting flooring for a foodservice facility, the ease of maintenance and cleaning is an essential consideration. While color, design, cost, and what other businesses use may be relevant factors, the primary concern should be ensuring the flooring can be kept clean and sanitary.

155. Which of the following should be included in a foodservice facility's waste disposal area design?
a) A designated area for storing soiled linens and uniforms
b) A direct connection between the waste disposal area and the kitchen
c) Waste receptacles placed near customer seating areas for convenience
d) Adequate space for separating recyclables, compostables, and garbage

Answer: d) Adequate space for separating recyclables, compostables, and garbage
Explanation: A foodservice facility's waste disposal area should include adequate space for separating recyclables, compostables, and garbage to promote proper waste management. Storing soiled linens, connecting the waste area to the kitchen, or placing receptacles near customer seating would not be appropriate choices.

156. What is the primary goal of proper ventilation in a foodservice facility?
a) To prevent condensation on windows
b) To eliminate or reduce cooking odors
c) To control humidity and temperature for customer comfort
d) To remove heat, steam, and airborne contaminants from the cooking area

Answer: d) To remove heat, steam, and airborne contaminants from the cooking area
Explanation: The primary goal of proper ventilation in a foodservice facility is to remove heat, steam, and airborne contaminants from the cooking area, promoting a safe and sanitary work environment. While preventing condensation, reducing odors, and controlling temperature and humidity for customer comfort are important, they are secondary to maintaining a safe cooking area.

157. Which type of material is best suited for cutting boards in a foodservice facility?
a) Wood
b) Glass
c) Plastic
d) Marble

Answer: c) Plastic
Explanation: Plastic cutting boards are the best choice for foodservice facilities, as they are easy to clean and sanitize. Wood can be porous and harbor bacteria, while glass and marble can damage knives and lead to cross-contamination.

158. What is a crucial feature for foodservice equipment to ensure easy cleaning and maintenance?
a) Aesthetically pleasing design
b) Removable parts for cleaning
c) Expensive, high-quality materials
d) The latest technology advancements

Answer: b) Removable parts for cleaning
Explanation: Removable parts for cleaning are essential for foodservice equipment to ensure easy cleaning and maintenance. While aesthetics, quality materials, and technology advancements are important, they should not compromise the ability to maintain cleanliness and sanitation.

159. In a foodservice facility, which type of lighting is best suited for food preparation areas?
a) Dimmed, warm-colored lighting
b) Bright, natural-colored lighting
c) Colored lighting to match the facility's theme
d) Fluorescent lighting to save energy

Answer: b) Bright, natural-colored lighting
Explanation: Bright, natural-colored lighting is best suited for food preparation areas in a foodservice facility. This type of lighting allows staff to see the condition of the food and work area clearly, ensuring proper food handling and sanitation.

160. Which of the following materials is most appropriate for food-contact surfaces in a foodservice facility?
a) Porous materials, such as wood
b) Smooth, nonabsorbent materials, such as stainless steel
c) Absorbent materials, such as fabric or sponge
d) Materials with small cracks and crevices, such as ceramic

Answer: b) Smooth, nonabsorbent materials, such as stainless steel
Explanation: Smooth, nonabsorbent materials like stainless steel are most appropriate for food-contact surfaces in a foodservice facility. These surfaces are easy to clean and sanitize and do not harbor bacteria, unlike porous or absorbent materials or materials with small cracks and crevices.

161. Which factor is most important when designing a handwashing station in a foodservice facility?
a) Convenient location for customers
b) Proximity to food preparation areas and employee traffic
c) The inclusion of scented soap and lotion
d) A design that matches the overall decor of the facility

Answer: b) Proximity to food preparation areas and employee traffic
Explanation: The most important factor when designing a handwashing station in a foodservice facility is its proximity to food preparation areas and employee traffic. This encourages frequent and proper handwashing, which is essential for maintaining food safety. While convenience for customers, scented soap, and decor may be considered, they should not compromise the primary goal of promoting handwashing among employees.

162. Which of the following is the primary reason for proper ventilation in a foodservice facility?
a) To maintain a comfortable working temperature for employees
b) To control odors and remove excess moisture
c) To enhance the overall aesthetic of the facility
d) To prevent pests from entering the facility

Answer: b) To control odors and remove excess moisture
Explanation: Proper ventilation is crucial for controlling odors and removing excess moisture in a foodservice facility. While maintaining a comfortable working temperature, enhancing aesthetics, and preventing pests are important, the primary purpose of ventilation is to ensure a sanitary environment and reduce the risk of food contamination.

163. What is the recommended lighting intensity for food preparation areas in a foodservice facility?
a) 10 foot-candles (108 lux)
b) 20 foot-candles (215 lux)
c) 50 foot-candles (538 lux)
d) 100 foot-candles (1076 lux)

Answer: c) 50 foot-candles (538 lux)
Explanation: The recommended lighting intensity for food preparation areas in a foodservice facility is 50 foot-candles (538 lux). This level of lighting ensures that employees can clearly see the condition of the food and their work area, which is vital for maintaining food safety and sanitation.

164. Which type of waste disposal container should be used for disposing of grease and oil in a foodservice facility?
a) Standard trash can with a liner
b) Recyclable container designated for grease and oil
c) Grease interceptor or trap
d) Compost bin

Answer: c) Grease interceptor or trap
Explanation: A grease interceptor or trap should be used for disposing of grease and oil in a foodservice facility. This specialized container prevents grease and oil from entering the sewer system, reducing the risk of blockages and environmental contamination.

165. What is a crucial step in managing waste disposal in a foodservice facility?
a) Disposing of waste in the nearest outdoor dumpster
b) Separating waste into different categories for proper disposal
c) Ignoring the need for a waste disposal schedule
d) Storing waste in the same area as food and equipment

Answer: b) Separating waste into different categories for proper disposal
Explanation: Separating waste into different categories for proper disposal is a crucial step in managing waste disposal in a foodservice facility. This ensures that each type of waste is handled and disposed of correctly, which helps maintain cleanliness, reduce odors, and minimize the risk of pest infestations.

166. Which of the following is an essential aspect of proper waste disposal in a foodservice facility?
a) Keeping waste containers open for easy access
b) Storing waste containers near food preparation areas
c) Using tight-fitting lids and maintaining cleanliness around waste containers
d) Disposing of waste only when the containers are completely full

Answer: c) Using tight-fitting lids and maintaining cleanliness around waste containers
Explanation: Using tight-fitting lids and maintaining cleanliness around waste containers is essential for proper waste disposal in a foodservice facility. This practice helps to control odors, minimize the risk of cross-contamination, and deter pests.

167. Which of the following is the primary purpose of a backflow prevention device in a foodservice facility?
a) To regulate water pressure
b) To prevent the reverse flow of contaminated water into the potable water supply
c) To remove contaminants from the water
d) To measure water usage

Answer: b) To prevent the reverse flow of contaminated water into the potable water supply
Explanation: The primary purpose of a backflow prevention device is to prevent the reverse flow of contaminated water into the potable water supply. This helps protect the water supply from contamination, ensuring that safe water is available for food preparation and cleaning.

168. What is the minimum water temperature required for handwashing in a foodservice facility?
a) 70°F (21°C)
b) 85°F (29°C)
c) 100°F (38°C)
d) 120°F (49°C)

Answer: c) 100°F (38°C)
Explanation: The minimum water temperature required for handwashing in a foodservice facility is 100°F (38°C). This temperature is effective in removing dirt, grease, and pathogens while still being comfortable for employees to use regularly.

169. Which of the following is an essential component of proper plumbing design in a foodservice facility?
a) Installing floor drains directly beneath equipment
b) Using horizontal pipes for drainage
c) Providing air gaps between the water supply inlet and the flood level of a plumbing fixture
d) Connecting the waste line of a dishwasher directly to the sewer line

Answer: c) Providing air gaps between the water supply inlet and the flood level of a plumbing fixture
Explanation: Providing air gaps between the water supply inlet and the flood level of a plumbing fixture is an essential component of proper plumbing design in a foodservice facility. Air gaps help prevent backflow and ensure that contaminated water cannot enter the potable water supply.

170. Which of the following is a critical aspect of maintaining a safe water supply in a foodservice facility?
a) Regularly testing water for the presence of pathogens
b) Ignoring the need for water treatment systems
c) Using untreated well water for food preparation
d) Allowing direct connections between the potable water supply and the wastewater system

Answer: a) Regularly testing water for the presence of pathogens
Explanation: Regularly testing water for the presence of pathogens is a critical aspect of maintaining a safe water supply in a foodservice facility. This practice helps ensure that the water used for food preparation and cleaning meets safety standards and reduces the risk of foodborne illness.

171. Which type of backflow prevention device is required for high-hazard situations, such as when chemicals are used in the water system?
a) Vacuum breaker
b) Air gap
c) Double check valve assembly
d) Reduced pressure zone (RPZ) device

Answer: d) Reduced pressure zone (RPZ) device
Explanation: A reduced pressure zone (RPZ) device is required for high-hazard situations, such as when chemicals are used in the water system. This type of backflow prevention device provides the highest level of protection against backflow and is designed to protect potable water supplies from contamination.

172. Which of the following is the best method for calibrating a bimetallic stemmed thermometer?
a) Placing the thermometer in boiling water and adjusting the calibration nut
b) Immersing the thermometer in a glass of water and adjusting the calibration nut
c) Placing the thermometer in an ice slurry and adjusting the calibration nut
d) Immersing the thermometer in hot oil and adjusting the calibration nut

Answer: c) Placing the thermometer in an ice slurry and adjusting the calibration nut
Explanation: The best method for calibrating a bimetallic stemmed thermometer is to place it in an ice slurry and adjust the calibration nut until it reads 32°F (0°C). This ensures accurate temperature readings at the critical freezing point.

173. What is the primary reason for regularly calibrating thermometers in a foodservice establishment?
a) To extend the life of the thermometer
b) To ensure accurate temperature readings
c) To comply with manufacturer warranties
d) To keep track of the thermometer's age

Answer: b) To ensure accurate temperature readings. Explanation: The primary reason for regularly calibrating thermometers in a foodservice establishment is to ensure accurate temperature readings. This helps maintain food safety by ensuring proper cooking, cooling, and holding temperatures.

174. Which of the following should be checked regularly to maintain the accuracy of a digital thermometer?
a) Battery life
b) Calibration nut
c) Sensitivity of the probe
d) Length of the stem

Answer: a) Battery life
Explanation: Regularly checking the battery life of a digital thermometer is essential for maintaining its accuracy. A low battery can cause inaccurate temperature readings, leading to potential food safety issues.

175. What is the recommended frequency for calibrating thermometers in a foodservice establishment?
a) Once a year
b) Once a month
c) Once a week
d) Daily

Answer: c) Once a week
Explanation: It is recommended to calibrate thermometers in a foodservice establishment once a week to ensure accurate temperature readings. However, more frequent calibration may be necessary if the thermometer is subjected to extreme temperature changes or physical impacts.

176. Which of the following is true regarding the calibration of infrared thermometers?
a) Infrared thermometers do not require calibration
b) Calibration can only be done by the manufacturer
c) Calibration should be done using a standard reference surface
d) Calibration can be done using an ice slurry

Answer: c) Calibration should be done using a standard reference surface
Explanation: Infrared thermometers should be calibrated using a standard reference surface, such as a blackbody calibrator, to ensure accurate temperature readings. Unlike bimetallic stemmed thermometers, they cannot be calibrated using an ice slurry or boiling water.

177. In the event of a power outage, what should be done with perishable food in a foodservice establishment?
a) Immediately discard all perishable food
b) Keep the refrigerator and freezer doors closed as much as possible
c) Relocate the food to a different facility with power
d) Cook and serve the perishable food immediately

Answer: b) Keep the refrigerator and freezer doors closed as much as possible
Explanation: During a power outage, keeping the refrigerator and freezer doors closed as much as possible helps maintain the temperature and preserve perishable food for a longer period. Food safety procedures should be followed to determine if any food needs to be discarded based on temperature and the duration of the power outage.

178. Which of the following is the primary purpose of a food recall?
a) To remove potentially hazardous food from the market
b) To protect the reputation of the food manufacturer
c) To comply with government regulations
d) To ensure a competitive advantage for the foodservice establishment

Answer: a) To remove potentially hazardous food from the market
Explanation: The primary purpose of a food recall is to remove potentially hazardous food from the market, protecting public health by preventing foodborne illnesses and other adverse health effects.

179. When a food recall is initiated, what is the first step a foodservice establishment should take?
a) Discard all affected products immediately
b) Inform customers about the recall
c) Identify and isolate affected products
d) Contact the local health department

Answer: c) Identify and isolate affected products
Explanation: The first step a foodservice establishment should take when a food recall is initiated is to identify and isolate affected products to prevent their use or distribution. Further actions, such as discarding the products or returning them to the supplier, will be determined based on the recall notice's instructions.

180. What information should be documented during a food recall process?
a) The names and contact information of all customers who consumed the affected product
b) The amount and type of affected product, date received, and supplier information
c) The financial impact of the recall on the foodservice establishment
d) The personal opinions of staff members regarding the recall

Answer: b) The amount and type of affected product, date received, and supplier information
Explanation: During a food recall process, it is crucial to document the amount and type of affected product, date received, and supplier information. This documentation is necessary for traceability and may be required by regulatory authorities or suppliers to verify that the recall has been appropriately executed.

181. What is the appropriate course of action for a foodservice establishment if a customer experiences an allergic reaction while dining?
a) Ask the customer to leave the premises immediately
b) Offer the customer an antihistamine medication
c) Call emergency medical services (EMS) if the reaction is severe
d) Ignore the situation and continue serving other customers

Answer: c) Call emergency medical services (EMS) if the reaction is severe
Explanation: If a customer experiences a severe allergic reaction while dining, the appropriate course of action is to call emergency medical services (EMS). Staff should be trained to recognize the signs of an allergic reaction and know how to respond effectively to protect the customer's health and safety.

182. Which of the following best describes the role of a food safety manager in a foodservice establishment?
a) To oversee daily food preparation and service, ensuring quality and customer satisfaction
b) To ensure compliance with all applicable food safety regulations and maintain a sanitary environment
c) To solely manage the financial aspects of the foodservice establishment
d) To create and implement marketing campaigns to attract new customers

Answer: b) To ensure compliance with all applicable food safety regulations and maintain a sanitary environment
Explanation: The primary role of a food safety manager in a foodservice establishment is to ensure compliance with all applicable food safety regulations and maintain a sanitary environment to protect public health and prevent foodborne illnesses.

183. Which of the following is the most appropriate action for a food safety manager when a foodborne illness outbreak is suspected?
a) Ignore the situation and hope it resolves on its own
b) Immediately close the establishment and fire all staff members
c) Contact the local health department and cooperate with the investigation
d) Post a public apology on social media and offer discounts to customers

Answer: c) Contact the local health department and cooperate with the investigation
Explanation: When a foodborne illness outbreak is suspected, the most appropriate action for a food safety manager is to contact the local health department and cooperate with the investigation. This ensures that the source of the outbreak is identified, and corrective actions are taken to protect public health.

184. What is the primary purpose of food safety regulations in the foodservice industry?
a) To create a competitive advantage for certain establishments
b) To protect public health and prevent foodborne illnesses
c) To increase government revenue through fines and penalties
d) To generate publicity for foodservice establishments

Answer: b) To protect public health and prevent foodborne illnesses
Explanation: The primary purpose of food safety regulations in the foodservice industry is to protect public health and prevent foodborne illnesses by establishing standards and guidelines for safe food handling and sanitation practices.

185. Which of the following is NOT a responsibility of a food safety manager?
a) Ensuring that all staff members receive proper food safety training
b) Maintaining and updating the establishment's food safety plan
c) Personally cooking and serving all meals
d) Regularly monitoring and documenting food temperatures and sanitation practices

Answer: c) Personally cooking and serving all meals
Explanation: While a food safety manager is responsible for ensuring compliance with food safety regulations, providing staff training, and maintaining the establishment's food safety plan, their role does not require them to personally cook and serve all meals. Their focus should be on overseeing and managing the food safety program.

186. Which government agency is primarily responsible for regulating food safety in the United States?
a) The Environmental Protection Agency (EPA)
b) The Federal Communications Commission (FCC)
c) The Food and Drug Administration (FDA)
d) The Department of Housing and Urban Development (HUD)

Answer: c) The Food and Drug Administration (FDA)
Explanation: The Food and Drug Administration (FDA) is the primary government agency responsible for regulating food safety in the United States. They work in conjunction with other agencies, such as the United States Department of Agriculture (USDA), to establish and enforce food safety regulations and guidelines.

187. Which organization is responsible for inspecting meat, poultry, and egg products in the United States?
a) The Food and Drug Administration (FDA)
b) The Environmental Protection Agency (EPA)
c) The United States Department of Agriculture (USDA)
d) The Centers for Disease Control and Prevention (CDC)

Answer: c) The United States Department of Agriculture (USDA)
Explanation: The USDA is responsible for inspecting meat, poultry, and egg products in the United States, ensuring they are safe and properly labeled. The FDA oversees food safety for other types of food products.

188. What is the purpose of the Food Code, a model set of guidelines published by the FDA?
a) To provide legal requirements for all foodservice establishments in the United States
b) To serve as a recommendation for state and local governments when developing food safety regulations
c) To establish marketing standards for the food industry
d) To dictate the nutritional content of all food products

Answer: b) To serve as a recommendation for state and local governments when developing food safety regulations
Explanation: The Food Code is a model set of guidelines published by the FDA that serves as a recommendation for state and local governments when developing their own food safety regulations. It is not a legal requirement but serves as a basis for many state and local food safety laws.

189. Which federal agency is responsible for investigating and responding to foodborne illness outbreaks in the United States?
a) The Food and Drug Administration (FDA)
b) The Centers for Disease Control and Prevention (CDC)
c) The United States Department of Agriculture (USDA)
d) The Environmental Protection Agency (EPA)

Answer: b) The Centers for Disease Control and Prevention (CDC). Explanation: The CDC is responsible for investigating and responding to foodborne illness outbreaks in the United States. They work closely with other agencies, such as the FDA and USDA, as well as state and local health departments to identify the source of outbreaks and implement control measures.

190. In addition to federal regulations, foodservice establishments must also comply with:
a) Only state regulations
b) Only local regulations
c) Both state and local regulations
d) Neither state nor local regulations

Answer: c) Both state and local regulations. Explanation: Foodservice establishments must comply with federal, state, and local regulations. State and local regulations may be more stringent than federal regulations and can vary depending on the jurisdiction.

191. Which of the following food safety regulations is typically enforced at the state or local level?
a) The Hazard Analysis Critical Control Point (HACCP) system
b) The Food Safety Modernization Act (FSMA)
c) Health department inspections and permits
d) The Nutrition Labeling and Education Act (NLEA)

Answer: c) Health department inspections and permits
Explanation: Health department inspections and permits are typically enforced at the state or local level. Inspections ensure that foodservice establishments comply with food safety regulations, and permits are often required to operate a foodservice establishment.

192. What is the primary responsibility of a food manager in a foodservice establishment?
a) Preparing and serving food
b) Ensuring profitability
c) Overseeing staff training and maintaining food safety
d) Developing new menu items

Answer: c) Overseeing staff training and maintaining food safety
Explanation: While a food manager may have many responsibilities, their primary responsibility is to oversee staff training and ensure food safety within the establishment. This includes implementing and monitoring food safety practices, training employees, and maintaining compliance with food safety regulations.

193. Which of the following is a key component of a food manager's role in promoting food safety?
a) Developing and implementing a Hazard Analysis Critical Control Point (HACCP) plan
b) Designing the menu and selecting ingredients
c) Managing the marketing and advertising for the establishment
d) Setting pricing and managing the budget

Answer: a) Developing and implementing a Hazard Analysis Critical Control Point (HACCP) plan
Explanation: A key component of a food manager's role in promoting food safety is developing and implementing a HACCP plan. This plan identifies potential hazards in the food production process and establishes critical control points to monitor and control those hazards, ensuring food safety.

194. Which certification is often required for food managers to demonstrate their knowledge of food safety?
a) ServSafe Food Manager Certification
b) Nutrition and Dietetics Certification
c) Culinary Arts Certification
d) Restaurant Management Certification

Answer: a) ServSafe Food Manager Certification
Explanation: The ServSafe Food Manager Certification is a widely recognized certification that demonstrates a food manager's knowledge of food safety. Many states and local jurisdictions require food managers to obtain this or a similar certification to ensure they are knowledgeable about food safety practices and regulations.

195. In addition to maintaining food safety, a food manager is also responsible for:
a) Setting the menu prices
b) Managing employee scheduling and payroll
c) Ensuring adequate staffing levels and training
d) Performing all cooking tasks in the establishment

Answer: c) Ensuring adequate staffing levels and training
Explanation: A food manager is responsible for ensuring adequate staffing levels and training. This includes hiring, training, and supervising employees to ensure they are knowledgeable about food safety and capable of performing their job duties effectively.

196. What is one of the most important aspects of a food manager's role in ensuring food safety compliance among employees?
a) Taking disciplinary actions against employees who violate food safety rules
b) Training employees on food safety procedures and monitoring their adherence to these practices
c) Personally preparing all food items to ensure they are safe for consumption
d) Designing the layout of the kitchen and dining areas

Answer: b) Training employees on food safety procedures and monitoring their adherence to these practices
Explanation: One of the most important aspects of a food manager's role in ensuring food safety compliance among employees is training them on food safety procedures and monitoring their adherence to these practices. This includes providing ongoing training and ensuring employees understand and follow food safety guidelines.

197. Which of the following records is most important for a foodservice establishment to maintain for food safety compliance?
a) Employee vacation schedules
b) Marketing materials and advertisements
c) Temperature logs for cooking, cooling, and reheating foods
d) Customer feedback forms

Answer: c) Temperature logs for cooking, cooling, and reheating foods
Explanation: Temperature logs for cooking, cooling, and reheating foods are essential records for maintaining food safety compliance. These logs help ensure that food is cooked, cooled, and reheated according to safety guidelines, reducing the risk of foodborne illness.

198. In the event of a foodborne illness outbreak, which documentation can help trace the source of the issue?
a) Employee training records
b) Financial records and invoices
c) Supplier delivery records and invoices
d) Customer satisfaction surveys

Answer: c) Supplier delivery records and invoices. Explanation: Supplier delivery records and invoices can help trace the source of a foodborne illness outbreak by identifying the origin of potentially contaminated ingredients. This information is crucial for identifying the cause of the outbreak and taking appropriate corrective actions.

199. How long should food safety records, such as temperature logs and cleaning schedules, be retained by a foodservice establishment?
a) One month
b) Six months
c) One year
d) As required by local regulations

Answer: d) As required by local regulations. Explanation: The retention period for food safety records, such as temperature logs and cleaning schedules, varies depending on local regulations. It is essential for foodservice establishments to be aware of and adhere to these requirements to ensure compliance.

200. Which of the following is an essential component of a food safety management system in a foodservice establishment?
a) Keeping a detailed inventory of all kitchen equipment
b) Monitoring social media accounts for customer feedback
c) Documenting employee performance reviews
d) Maintaining records of regular equipment maintenance and calibration

Answer: d) Maintaining records of regular equipment maintenance and calibration. Explanation: Maintaining records of regular equipment maintenance and calibration is an essential component of a food safety management system. This documentation ensures that equipment is functioning correctly and accurately measuring temperatures, helping to prevent food safety hazards.

201. Why is it important for a foodservice establishment to document employee food safety training?
a) To use as a marketing tool to attract customers
b) To comply with food safety regulations and demonstrate due diligence
c) To ensure that employees understand their job responsibilities
d) To provide a basis for employee promotions

Answer: b) To comply with food safety regulations and demonstrate due diligence
Explanation: Documenting employee food safety training is important for compliance with food safety regulations and demonstrating due diligence. These records show that a foodservice establishment is committed to maintaining a safe environment and properly training its staff on food safety practices.

202. During a food safety inspection, the inspector identifies a critical violation related to temperature control. What should the foodservice establishment do immediately?
a) Wait for the inspector's final report before taking any action
b) Make a note of the violation for future reference
c) Correct the violation as soon as possible
d) Notify customers of the violation

Answer: c) Correct the violation as soon as possible
Explanation: When a critical violation is identified during a food safety inspection, the foodservice establishment should take immediate corrective action to mitigate any potential risks to public health.

203. What is the primary purpose of a food safety inspection?
a) To penalize foodservice establishments for non-compliance
b) To ensure that foodservice establishments are following food safety regulations
c) To evaluate the quality of the food being served
d) To provide recommendations for improving the menu

Answer: b) To ensure that foodservice establishments are following food safety regulations
Explanation: The primary purpose of a food safety inspection is to ensure that foodservice establishments are following food safety regulations to protect public health and prevent foodborne illnesses.

204. Which of the following actions is most likely to result in a food safety violation?
a) Cleaning and sanitizing cutting boards between uses
b) Storing raw meat above ready-to-eat foods in the refrigerator
c) Ensuring proper handwashing practices among staff members
d) Monitoring and documenting food temperatures during cooking and cooling

Answer: b) Storing raw meat above ready-to-eat foods in the refrigerator
Explanation: Storing raw meat above ready-to-eat foods in the refrigerator can result in cross-contamination and is a food safety violation. Raw meat should be stored below ready-to-eat foods to prevent contamination.

205. In the event of a food safety violation, what is the most appropriate course of action for a foodservice establishment?
a) Disregard the violation and continue with regular operations
b) Correct the violation and implement measures to prevent future occurrences
c) Deny the violation and challenge the inspector's findings
d) Close the establishment until the issue is resolved

Answer: b) Correct the violation and implement measures to prevent future occurrences
Explanation: When a food safety violation is identified, the foodservice establishment should correct the violation and implement measures to prevent future occurrences. This demonstrates a commitment to food safety and compliance with regulations.

206. Which of the following factors is likely to have the most significant impact on the frequency of food safety inspections for a foodservice establishment?
a) The size of the establishment
b) The establishment's location
c) The type of food being served
d) The establishment's compliance history

Answer: d) The establishment's compliance history
Explanation: The frequency of food safety inspections is often determined by an establishment's compliance history. Establishments with a history of violations or non-compliance may be subject to more frequent inspections to ensure that they are adhering to food safety regulations.

207. What is the primary goal of a food safety management system, such as HACCP?
a) To improve the taste and presentation of food
b) To increase the profitability of a foodservice establishment
c) To ensure food is prepared and served in a safe and sanitary manner
d) To reduce the amount of food waste produced by a foodservice establishment

Answer: c) To ensure food is prepared and served in a safe and sanitary manner
Explanation: A food safety management system, such as HACCP, is designed to ensure food is prepared and served in a safe and sanitary manner, reducing the risk of foodborne illnesses.

208. Which of the following is NOT one of the seven principles of HACCP?
a) Conduct a hazard analysis
b) Implement staff training programs
c) Establish critical limits
d) Establish corrective actions

Answer: b) Implement staff training programs
Explanation: While staff training is essential for effective food safety management, it is not one of the seven principles of HACCP. The principles include conducting a hazard analysis, identifying critical control points, establishing critical limits, monitoring procedures, establishing corrective actions, verifying the system is working, and keeping records.

209. What is the purpose of identifying critical control points (CCPs) in a HACCP system?
a) To pinpoint the steps in the food preparation process where hazards can be effectively controlled
b) To determine the areas where staff training is most needed
c) To identify the most profitable menu items
d) To monitor food waste and improve efficiency

Answer: a) To pinpoint the steps in the food preparation process where hazards can be effectively controlled
Explanation: Identifying critical control points (CCPs) in a HACCP system helps to pinpoint the steps in the food preparation process where hazards can be effectively controlled, thereby reducing the risk of foodborne illnesses.

210. Which of the following is an example of a critical limit in a HACCP system?
a) Requiring all employees to wash their hands for at least 20 seconds
b) Cooking poultry to a minimum internal temperature of 165°F (74°C) for at least 15 seconds
c) Ensuring that all food handlers have completed a food safety training course
d) Discarding any food that has been stored at room temperature for more than two hours

Answer: b) Cooking poultry to a minimum internal temperature of 165°F (74°C) for at least 15 seconds. Explanation: A critical limit is a specific and measurable parameter, such as temperature or time, that must be met to ensure the control of a food safety hazard. Cooking poultry to a minimum internal temperature of 165°F (74°C) for at least 15 seconds is an example of a critical limit in a HACCP system.

211. In a HACCP system, what is the purpose of establishing monitoring procedures?
a) To determine the profitability of menu items
b) To ensure that critical control points are being consistently controlled
c) To evaluate the effectiveness of staff training programs
d) To track customer feedback and satisfaction

Answer: b) To ensure that critical control points are being consistently controlled

Explanation: Establishing monitoring procedures in a HACCP system ensures that critical control points are being consistently controlled, enabling the foodservice establishment to take corrective action if necessary and reduce the risk of foodborne illnesses.

As you conclude this study guide, take a moment to reflect on the wealth of information you've absorbed. From understanding the principles of food safety and temperature control to pest management, facility design, and the intricacies of food safety regulations, you've taken significant strides in preparing yourself for the exam. Along the way, we delved into the responsibilities of food managers, the importance of proper record-keeping, and the steps needed to address inspections, violations, and corrective actions. By studying these essential topics, you have built a solid foundation for implementing and managing food safety systems like HACCP.

Remember, everyone faces challenges and setbacks, but it's your ability to learn from these experiences and move forward that defines your growth. In the world of food safety, you have the power to protect the health and well-being of countless individuals, and the knowledge you've gained from this study guide will help you make a meaningful impact.

As you continue on your journey, never forget that your dreams are valid, your fears can be overcome, and your concerns can be transformed into confidence through diligence and perseverance. Carry the knowledge you've gained here with pride and embrace the opportunities that lie ahead.

In conclusion, we want to wish you the very best of luck as you tackle the exam and embark on your career in food safety management. Remember to believe in yourself, stay focused, and keep striving for excellence. You have the power to make a difference, and we have no doubt that you will rise to the occasion.

Printed in the USA
CPSIA information can be obtained
at www.ICGtesting.com
LVHW080540061223
765518LV00079B/2105